*A Lifetime of Synergy
with Theory and Experiment*

# A Lifetime of Synergy with Theory and Experiment

Andrew Streitwieser

PROFILES, PATHWAYS, AND DREAMS
Autobiographies of Eminent Chemists

*Jeffrey I. Seeman, Series Editor*

American Chemical Society, Washington, DC

Library of Congress Cataloging-in-Publication Data

Streitwieser, Andrew, 1927–
  A lifetime of synergy with theory and experiment / Andrew Streitwieser

  p. cm.—(Profiles, pathways, and dreams, ISSN 1047–8329)
  Includes bibliographical references (p.   -   ) and index.

  ISBN 0–8412–1836–6 (case: alk. paper)

  1. Streitwieser, Andrew, 1927–    . 2. Chemists—United States—Biography.
  3. Chemistry, Organic—United States—History—20th century.

  I. Title. II. Series.

QD22.S78S77  1996
540'.92—dc20
[B]                                                              96–27017
                                                                    CIP

Jeffrey I. Seeman, Series Editor

The paper used in this publication meets the minimum requirements of American National Standard for Information Sciences—Permanence of Paper for Printed Library Materials, ANSI Z39.48–1984.

Copyright © 1997

American Chemical Society

All Rights Reserved. The copyright owner consents that reprographic copies may be made for personal or internal use or for the personal or internal use of specific clients. This consent is given on the condition, however, that the copier pay the stated per-copy fee through the Copyright Clearance Center, Inc., 222 Rosewood Drive, Danvers, MA 01923, for copying beyond that permitted by Sections 107 or 108 of the U.S. Copyright Law. This consent does not extend to copying or transmission by any means—graphic or electronic—for any other purpose, such as for general distribution, for advertising or promotional purposes, for creating a new collective work, for resale, or for information storage and retrieval systems. The copying fee is $4.00 per page. Please report your copying to the Copyright Clearance Center with this code: 1047–8329/96/$00.00+4.00.

The citation of trade names and/or names of manufacturers in this publication is not to be construed as an endorsement or as approval by ACS of the commercial products or services referenced herein; nor should the mere reference herein to any drawing, specification, chemical process, or other data be regarded as a license or as a conveyance of any right or permission to the holder, reader, or any other person or corporation, to manufacture, reproduce, use, or sell any patented invention or copyrighted work that may in any way be related thereto. Registered names, trademarks, etc., used in this publication, even without specific indication thereof, are not to be considered unprotected by law.

PRINTED IN THE UNITED STATES OF AMERICA

## ACS Books Advisory Board

Robert J. Alaimo
Procter & Gamble Pharmaceuticals

Mark Arnold
University of Iowa

David Baker
University of Tennessee

Arindam Bose
Pfizer Central Research

Robert F. Brady, Jr.
Naval Research Laboratory

Mary E. Castellion
ChemEdit Company

Margaret A. Cavanaugh
National Science Foundation

Arthur B. Ellis
University of Wisconsin at Madison

Gunda I. Georg
University of Kansas

Madeleine M. Joullie
University of Pennsylvania

Lawrence P. Klemann
Nabisco Foods Group

Douglas R. Lloyd
The University of Texas at Austin

Cynthia A. Maryanoff
R. W. Johnson Pharmaceutical
   Research Institute

Roger A. Minear
University of Illinois
   at Urbana–Champaign

Omkaram Nalamasu
AT&T Bell Laboratories

Vincent Pecoraro
University of Michigan

George W. Roberts
North Carolina State University

John R. Shapley
University of Illinois
   at Urbana–Champaign

Douglas A. Smith
Concurrent Technologies Corporation

L. Somasundaram
DuPont

Michael D. Taylor
Parke-Davis Pharmaceutical Research

William C. Walker
DuPont

Peter Willett
University of Sheffield (England)

# Foreword

In 1986, the ACS Books Department accepted for publication a collection of autobiographies of organic chemists, to be published in a single volume. However, the authors were much more prolific than the project's editor, Jeffrey I. Seeman, had anticipated, and under his guidance and encouragement, the project took on a life of its own. The original volume evolved into 22 volumes, and the first volume of Profiles, Pathways, and Dreams: Autobiographies of Eminent Chemists was published in 1990. Unlike the original volume, the series was structured to include chemical scientists in all specialties, not just organic chemistry. Our hope is that those who know the authors will be confirmed in their admiration for them, and that those who do not know them will find these eminent scientists a source of inspiration and encouragement, not only in any scientific endeavors, but also in life.

# *Contributors*

We thank the following corporations and Herchel Smith for their generous financial support of the series Profiles, Pathways, and Dreams.

Akzo nv

Bachem Inc.

DuPont

Duphar B.V.

Eisai Co., Ltd.

Fujisawa Pharmaceutical Co., Ltd.

Hoechst Celanese Corporation

Imperial Chemical Industries PLC

Kao Corporation

Mitsui Petrochemical Industries, Ltd.

The NutraSweet Company

Organon International B.V.

Pergamon Press PLC

Pfizer Inc.

Philip Morris

Quest International

Sandoz Pharmaceuticals Corporation

Sankyo Company, Ltd.

Schering–Plough Corporation

Shionogi Research Laboratories, Shionogi & Co., Ltd.

Herchel Smith

Suntory Institute for Bioorganic Research

Takasago International Corporation

Takeda Chemical Industries, Ltd.

Unilever Research U.S., Inc.

# Profiles, Pathways, and Dreams

## Titles in This Series

**Sir Derek H. R. Barton**  *Some Recollections of Gap Jumping*

**Arthur J. Birch**  *To See the Obvious*

**Melvin Calvin**  *Following the Trail of Light: A Scientific Odyssey*

**Donald J. Cram**  *From Design to Discovery*

**Michael J. S. Dewar**  *A Semiempirical Life*

**Carl Djerassi**  *Steroids Made It Possible*

**Ernest L. Eliel**  *From Cologne to Chapel Hill*

**Egbert Havinga**  *Enjoying Organic Chemistry, 1927–1987*

**Rolf Huisgen**  *The Adventure Playground of Mechanisms and Novel Reactions*

**William S. Johnson**  *A Fifty-Year Love Affair with Organic Chemistry*

**Raymond U. Lemieux**  *Explorations with Sugars: How Sweet It Was*

**Herman Mark**  *From Small Organic Molecules to Large: A Century of Progress*

**Bruce Merrifield**  *Life During a Golden Age of Peptide Chemistry: The Concept and Development of Solid-Phase Peptide Synthesis*

**Teruaki Mukaiyama**  *To Catch the Interesting While Running*

**Koji Nakanishi**  *A Wandering Natural Products Chemist*

**Tetsuo Nozoe**  *Seventy Years in Organic Chemistry*

**Vladimir Prelog**  *My 132 Semesters of Chemistry Studies*

**John D. Roberts**  *The Right Place at the Right Time*

**Paul von Rague Schleyer**  *From the Ivy League into the Honey Pot*

**F. Gordon A. Stone**  *Leaving No Stone Unturned: Pathways in Organometallic Chemistry*

**Andrew Streitwieser, Jr.**  *A Lifetime of Synergy with Theory and Experiment*

**Cheves Walling**  *Fifty Years of Free Radicals*

# *About the Editor*

JEFFREY I. SEEMAN received his B.S. with high honors in 1967 from the Stevens Institute of Technology in Hoboken, New Jersey, and his Ph.D. in organic chemistry in 1971 from the University of California, Berkeley. Following a two-year staff fellowship at the Laboratory of Chemical Physics of the National Institutes of Health in Bethesda, Maryland, he joined the Philip Morris Research Center in Richmond, Virginia. In 1983–1984, he enjoyed a sabbatical year at the Dyson Perrins Laboratory in Oxford, England, and claims to have visited more than 90% of the castles in England, Wales, and Scotland.

Seeman's 100 published papers, patents, and books include research in the areas of photochemistry, nicotine and tobacco alkaloid chemistry and synthesis, conformational analysis, pyrolysis chemistry, organotransition metal chemistry, the use of cyclodextrins for chiral recognition, and structure–activity relationships in olfaction. He was a plenary lecturer at the Eighth IUPAC Conference on Physical Organic Chemistry and has been an invited lecturer at numerous scientific meetings and universities. From 1989 to 1994, Seeman served on the Petroleum Research Fund Advisory Board, and he currently serves on the advisory board of the *Journal of Organic Chemistry*. Seeman continues to count Nero Wolfe and Archie Goodwin among his best friends.

# Contents

| | |
|---|---|
| List of Photographs | xiii |
| Preface | xix |
| Editor's Note | xxiii |
| About the Writing of This Book | 1 |
| Before College | 5 |
|     Early Years and Organic Specialties | 5 |
|     My First Paper | 17 |
|     Science Talent Search | 20 |
| Higher Education and the Army | 23 |
|     Columbia College and Army Service, 1945–1948 | 23 |
|     Graduate School, 1948–1951 | 29 |
|     AEC Postdoctoral Fellowship, MIT, 1951–1952 | 39 |
| University of California, Berkeley | 45 |
|     Early Years | 45 |
|     Stereochemistry of the Primary Carbon | 48 |
|     Solvolytic Displacement Reactions | 58 |
|     Secondary Deuterium Isotope Effects | 63 |
|     Molecular Orbital Theory for Organic Chemists | 68 |

| | |
|---|---:|
| Carbon Acidity | 85 |
|    Kinetic Acidities in Cyclohexylamine | 85 |
|    Equilibrium Acidities in Cyclohexylamine | 97 |
|    Brønsted Correlations | 102 |
|    Consulting at DuPont | 107 |
|    Mary Ann and Suzanne | 109 |
|    Organofluorine Carbanions | 115 |
|    Carbanions in Tetrahydrofuran | 125 |
| Theoretical Chemistry | 145 |
|    Computers and Molecular Orbital Theory | 145 |
|    Ab Initio Quantum Organic Chemistry | 149 |
|    Electron Density Functions | 154 |
|    Carboxylate Resonance | 166 |
|    Transition States and Ion-Pair Reactions | 170 |
|    Future Trends | 176 |
| f-Orbital Organometallic Chemistry | 181 |
|    Uranocene | 181 |
|    Organolanthanide Chemistry | 200 |
|    Cerocene | 201 |
|    Structures and Hydrolysis | 216 |
| Heterocycle Polycations | 223 |
|    "Densely Charged Compounds" | 223 |
| Odds and Ends | 237 |
|    Organic Plasma Chemistry | 237 |
|    Textbook: *Introduction to Organic Chemistry* | 238 |
|    *Journal of Organic Chemistry* | 243 |
|    Ethics | 246 |
|    Awards | 248 |
|    Hobbies | 251 |
|    Coda | 257 |
| Appendix A: Colleagues and Associates | 267 |
| Appendix B: Chemical Genealogy | 275 |
| References | 277 |
| Index | 293 |

# *Photographs*

Our intrepid editor, Jeff Seeman, and me at the ACS meeting in Anaheim, March 1995........................................................................... 3

Ed Kosower and me in Stuyvesant High School, around 1944................................................................................................ 6

My father and me, late 1927 or early 1928............................................ 10

With my roadster in Buffalo, 1930......................................................... 11

My mother and me in Germany in 1930 when I was 3 years old.................................................................................................. 11

A school picture of me at the age of 6.................................................. 11

My parents, about 1946. .......................................................................... 12

My mother, Sophie, in 1988, at the age of 84....................................... 13

Burg Streitwiesen, a castle above the hamlet of Streitwieser in Austria, 1980........................................................................................ 14

Edward Kosower at MIT in 1946 with the President of MIT, Karl T. Compton..................................................................................... 15

| | |
|---|---|
| Setting up a demonstration of my project for the Westinghouse Science Talent Search, Washington, DC, March 1945. | 21 |
| A recent picture of my brother Bill and me. | 24 |
| In the U.S Army Medical Corps, at Fitzsimmons General Hospital, Denver, CO, 1947. | 25 |
| Ben Widom and me at an awards dinner in Las Vegas in 1982. | 28 |
| William von Eggers Doering at Yale. | 31 |
| Kenneth and Marge Wiberg in Vancouver, 1992. | 32 |
| Al Wolf in 1964. | 33 |
| Jerome A. Berson in 1949. | 34 |
| Richard W. Young in 1950. | 36 |
| Part of the Doering group about 1947–1948 on a picnic: Milton Farber, Jerry Berson, Mrs. (Bella) Berson, Mortimer Levitz, Al Wolf, Mrs. (Helga) Wolf. | 38 |
| Jack Roberts in 1975. | 40 |
| William H. Saunders, Jr., in 1953. | 41 |
| With my first wife, Mary Ann, walking on campus with our first baby, David, in early 1954. | 46 |
| At the old chemistry building in January, 1961: Andrew Streitwieser, James Cason, Ignacio Tinoco, and Donald Noyce. | 47 |
| At the Faculty Club, March, 1983: William Dauben, Melvin Calvin, Don Noyce, Fritz Jensen, Andrew Streitwieser, and Paul Bartlett. | 48 |
| Liane Reif (now Reif-Lehrer), about 1959. | 56 |
| Bill Dauben, at a surprise birthday party, about 1970. | 59 |

| | |
|---|---|
| Professor Michael J. S. Dewar. | 72 |
| My children, David, 6, and Susan, 4, in 1960. | 74 |
| Bill Johnson and Edith and Jack Roberts at Fisher Island, Miami, April 1989. | 75 |
| Kenichi Fukui at the Sterling Winery, in October, 1972. | 79 |
| Presenting the Roger Adams Award to Georg Wittig during the National Organic Symposium, Tallahassee, FL, in June, 1973. | 80 |
| Shuji Ozawa, Sue, and me at Hakone Park with an awesome view of Mt. Fuji in October, 1971. | 81 |
| Sue and me and Professor Yukawa during my first trip to Japan in 1971. | 82 |
| Professors Tsuno and Yukawa during KISPOC I in 1982. | 83 |
| Posing in one of my laboratories in our brand-new Latimer Hall in May, 1963. | 86 |
| Heinz Koch in Grimentz, Switzerland, in 1972. | 88 |
| Professor Taniguchi with Heinz Koch during KISPOC V at Kyushi University in Fukuoka in 1993. | 89 |
| Bill Langworthy in his Berkeley lab in August, 1961. | 90 |
| Dale Van Sickle, Bill Langworthy, Heinz Koch, and John Brauman, at a party at the Koch's in 1960. | 91 |
| Playing ball at a group picnic in the 1960s. | 93 |
| Giving a seminar at the Universal Oil Products Company in 1963. | 98 |
| With my first wife, Mary Ann, in the apartment of Jack and Edith Roberts in early 1952. | 110 |
| My first wife, Mary Ann, and me on the occasion of my first award, in 1964. | 111 |

| | |
|---|---|
| My second wife, Sue, in 1966 before our engagement. | 114 |
| My children, Susan and David, in March, 1977, at our Berkeley home. | 116 |
| J. Colin Tatlow, Professor Emeritus, Birmingham University. | 117 |
| A birthday present from members of the research group in 1965: Herschel Rabitz, Dave Holtz, Bob Bittman, and Mark Bixler. | 119 |
| Speakers at a Conference on Carbanion Chemistry, Ottawa, Ontario, Canada, July, 1989: Laren Tolbert, Paul Schleyer, Dave Collum, Marye Anne Fox, Charles DePuy, Andrew Streitwieser, and Ted Cohen. | 142 |
| John Brauman playing the banjo at a group party in 1961. | 147 |
| Professor Charles Coulson. | 148 |
| John Pople and Henry F. (Fritz) Schaefer, III, at dinner in 1995. | 151 |
| A. Streitwieser, Peter Stang, Kurt Mislow, and Paul Schleyer, at a Symposium in Erlangen, Germany, April, 1990. | 156 |
| Richard F. W. Bader giving a talk in 1990. | 158 |
| The group designed this "Andyland" T-shirt and presented it to me at one of my group meetings in 1988. | 179 |
| Discussing chemistry with Professor Emanuel Vogel in Cologne in 1991. | 184 |
| With Ken Raymond at a party in 1978. | 185 |
| Speakers at a Natick Conference in October, 1969: Ernest Eliel, Alan W. Johnson, Duilio Arigoni, A. Streitwieser, Louis Long, Jr., Roald Hoffmann, Max Muxfeld, and Siegfried Hunig. | 187 |
| Sue with Jenny Green and Wilhelm Maier at lunch in the Napa Valley, California, in 1981. | 190 |

My inorganic colleague Richard A. Andersen at the Boca Verita (Mouth of Truth). ... 196

At the Second International Conference on Lanthanides and Actinides, Lisbon, April, 1987: Notker Rösch, Andrew Streitwieser, Patricia Watson, Norman Edelstein, and Hsu-Kun Wang. ... 199

With "eine Mass" (a one-liter container) of Munich beer. ... 203

Professor and Mrs. Ivar Ugi and their dog in 1991. ... 204

Johnny and Ullie Gasteiger at a dinner in 1995. ... 205

Sue and me in sheepskin coats in preparation for my 1976 Sabbatical in Germany. ... 206

Dinner at a restaurant near the University of Munich in 1991: Professors Knorr, Huisgen, and Gompper, my wife Sue, and Johnnie Gasteiger. ... 207

Heinrich Pfeiffer with Sue and me, December, 1978. ... 208

Professor Paul D. Bartlett on my right at a meeting of the Humboldt Foundation 1976. ... 209

Professor Ludwig Hofacker at the Technical University, Munich, in 1991. ... 210

In the new chemistry building of the Technical University, Munich, 1977. ... 211

With Frau Ingrid Braun, in 1985. ... 212

Professor Notker Rösch in 1978. ... 213

A portrait of me in my office in 1984. ... 220

At Ken Waterman's wedding celebration on the Berkeley campus, 1985: Leyi Gong, Scott Gronert, John Rigsbee, Dave Eisenberg, Mike Kaufman, Drew Speer, Phil Sasse, Steve Bachrach, Bob Moore, Andy Koch, and Jin-Xiang Ni. ... 229

Clayton Heathcock and me at Island Park, Idaho, in the summer of 1970. ... 239

| | |
|---|---|
| Lynne Gloria in 1968. | 241 |
| As Captain Kirk, repelling a group of "alien invaders" in a 1985 skit. | 242 |
| Paul Schleyer and Norman (Lou) Allinger during an editorial board meeting in 1994. | 246 |
| Receiving the 1967 American Chemical Society Award in Petroleum Chemistry from Chet Warner. | 248 |
| Lecturing in Israel in October of 1982. | 250 |
| A picture of me taken on the South America tour I took in 1994 to see the solar eclipse. | 252 |
| Two views of the eclipse of November 3, 1994. | 253 |
| Taken in Gubbio, Italy, *Gubbian Lock* was a prize-winning slide in a competition at the Berkeley Camera Club. | 254 |
| Emmett Eiland and David Shirley, at a wine tasting in Emmett's oriental rug store in 1985. | 256 |
| A rafting trip on the Stanislaus River in Northern California in the summer of 1982: my daughter Susan, Susan Johntz, our guide, Matt Lyttle, Dan Bors, A. Streitwieser, Scott Gronert, and Mike Kaufman. | 259 |
| With Professor Duilio Arigoni at the ETH, Zurich, in 1991. | 261 |
| With Vladimir Prelog in his office at the ETH, Zurich, in 1991. | 262 |
| Sue and me in our Munich apartment in 1991. | 263 |
| At a dinner in honor of my 65th birthday, in San Francisco: Kennith Smith, John Collins, Eusebio Juaristi, Spiro Alexandratos, and Bill Schriver. | 265 |

# *Preface*

"HOW DID YOU GET THE IDEA—and the good fortune—to convince 22 world-famous chemists to write their autobiographies?" This question has been asked of me, in these or similar words, frequently over the past several years. I hope to explain in this preface how the project came about, how the contributors were chosen, what the editorial ground rules were, what was the editorial context in which these scientists wrote their stories, and the answers to related issues. Furthermore, several authors specifically requested that the project's boundary conditions be known.

As I was preparing an article[1] for *Chemical Reviews* on the Curtin–Hammett principle, I became interested in the people who did the work and the human side of the scientific developments. I am a chemist, and I also have a deep appreciation of history, especially in the sense of individual accomplishments. Readers' responses to the historical section of that review encouraged me to take an active interest in the history of chemistry. The concept for Profiles, Pathways, and Dreams resulted from that interest.

My goal for Profiles was to document the development of modern organic chemistry by having individual chemists discuss their roles in this development. Authors were not chosen to represent my choice of the world's "best" organic chemists, as one might choose the "baseball all-star team of the century". Such an attempt would be foolish: Even the selection committees for the Nobel prizes do not make their decisions on such a premise.

The selection criteria were numerous. Each individual had to have made seminal contributions to organic chemistry over a multidecade career. (The average age of the authors is over 70!) Profiles would represent scientists born and professionally productive in different countries. (Chemistry in 13 countries is detailed.) Taken together, these individuals were to have conducted research in nearly all subspecialties of organic chemistry. Invitations to contribute were based on solicited advice and on recommendations of chemists from five continents, including nearly all of the contributors. The final assemblage was selected entirely and exclusively by me. Not all who were invited chose to participate, and not all who should have been invited could be asked.

A very detailed four-page document was sent to the contributors, in which they were informed that the objectives of the series were

1. to delineate the overall scientific development of organic chemistry during the past 30–40 years, a period during which this field has dramatically changed and matured;

2. to describe the development of specific areas of organic chemistry; to highlight the crucial discoveries and to examine the impact they have had on the continuing development in the field;

3. to focus attention on the research of some of the seminal contributors to organic chemistry; to indicate how their research programs progressed over a 20–40-year period; and

4. to provide a documented source for individuals interested in the hows and whys of the development of modern organic chemistry.

One noted scientist explained his refusal to contribute a volume by saying, in part, that "it is extraordinarily difficult to write in good taste about oneself. Only if one can manage a humorous and light touch does it come off well. Naturally, I would like to place my work in what I consider its true scientific perspective, but . . ."

Each autobiography reflects the author's science, his lifestyle, and the style of his research. Naturally, the volumes are not uniform, although each author attempted to follow the guidelines. "To write in good taste" was not an objective of the series. On the contrary, the authors were specifically requested not to write a review article of their field, but to detail their own research accomplishments. To the extent that this instruction was followed and the result is not "in good taste", then these are criticisms that I, as editor, must bear, not the writer.

As in any project, I have a few regrets. It is truly sad that Egbert Havinga and Herman Mark, who each wrote a volume, and David Ginsburg, who translated another, died during the course of this project. There have been many rewards, some of which are documented in my personal account of this project, entitled "Extracting the Essence: Adventures of an Editor" published in CHEMTECH.[2]

## Acknowledgments

I join the entire scientific community in offering each author unbounded thanks. I thank their families and their secretaries for their contributions. Furthermore, I thank numerous chemists for reading and reviewing the autobiographies, for lending photographs, for sharing information, and for providing each of the authors and me the encouragement to proceed in a project that was far more costly in time and energy than any of us had anticipated.

I thank my employer, Philip Morris USA, and J. Charles, R. N. Ferguson, K. Houghton, H. Grubbs, and W. F. Kuhn, for without their support Profiles, Pathways, and Dreams could not have been. I thank the staff of ACS Books for their hard work, dedication, and support. Each reader no doubt joins me in thanking 24 corporations and Herchel Smith for financial support for the project.

I thank my children, Jonathan and Brooke, for their patience and understanding; remarkably, I have been working on Profiles for more than half of their lives—probably the only half that they can remember! Finally, I again thank all those mentioned and especially my family, friends, colleagues, and the 22 authors for allowing me to share this experience with them.

JEFFREY I. SEEMAN
Philip Morris Research Center
Richmond, VA 23234

April 19, 1993

[1] Seeman, J. I. *Chem. Rev.* **1983**, *83*, 83–134.
[2] Seeman, J. I. *CHEMTECH* **1990**, *20*(2), 86–90.

# Editor's Note

VERY EARLY IN MY GRADUATE SCHOOL CAREER, I was frustrated to find that a particular volume of the *Journal of the American Chemical Society* was missing from the Berkeley chemistry library. The librarian recommended borrowing the issue from Professor Streitwieser, whose collection dated back to the 1940s. I soon found myself in Streitwieser's outer office. Lynne Gloria, his secretary (then, and for more than 25 years) said to go in and take what I needed. Hesitating, I knocked at the door. No answer. I knocked again. No answer. Lynne urged me forward. Upon entering, I noticed the professor working behind a pile of papers and books at his desk.

"Professor Streitwieser," I said, "I would like to borrow a journal."

No answer. I found the issue and thanked him. Silence. The same "exchange" was repeated a short time later, when I returned the journal. Impolite? No. His concentration was so intense, so focused on the chemistry he was studying, he simply did not realize I had entered, an intruder into his inner sanctum. I am especially amused by this encounter years later, now that I know one of his remarkable features is his definitive, emphatic, and resounding voice!

Intense focus and a distinctive voice are just two of the many facets of Andy Streitwieser. "The advent of the hot tub really changed him," concludes his daughter Susan, laughing. "After the hot tub, he became 'very Berkeley'. My favorite visual picture of Dad is in the hot tub, wearing his half glasses, completely absorbed in a chemistry journal!" Upon hearing this story, Sue, Streitwieser's wife, recently commented, "That certainly is the picture I too have of him, but he isn't always reading chemistry. Every evening, we converge at 5:00 p.m. in the hot tub to share a glass of wine and the events of the day. We often talk of plans, like future fishing trips! Sometimes we don't talk at all, just read."

As recounted in this autobiography, the suicide of Streitwieser's first wife had a great effect on him and his two children. Equally influential was his choosing Sue as his second wife. This was recently characterized by daughter Susan as "Sue marrying into the family. Dad looked for both a mate and a mother for us."

"I have learned many lessons in my life," Streitwieser mused recently. "I still have vivid memories of my first wife's death, and I remember the painful feelings of that period. But time really does heal. There are few things worth destroying one's life over. I remember other feelings of rejection, even those from grant proposals being turned down! I've learned not to overreact and, for example, to wait before responding to rejection. I also have learned over the years to listen without judging, especially when advising students."

When Streitwieser arrived in Berkeley in the early 1950s, the Chemistry Department certainly was a bastion of conservatism. Undoubtedly, this young, studious, and serious New Yorker fit well to this norm. As he tells it, "Growing up in New York, I became aware of a large slice of life, saw a lot at a young age. However, I never had a feeling of belonging to a community, of being a part of that location. You must get that when you are young; if you don't have that feeling as a child, you'll never get it." Yet, the sequence of photographs within his autobiography show the dynamics and influence of Northern California: Clean shaven in the 1960s, Streitwieser is bearded in the 1970s, and long-haired by the 1980s.

It is significant that Streitwieser's career as an organic chemist began as a collaborative effort, as a high school student in the early 1940s with two friends who would also later become professors of chemistry, E. M. Kosower and Lester Friedman. Few chemists who publish their first paper in the *Journal of the American Chemical Society* give their parents' home address and no other affiliation!

Because of an organic seminar requirement as a graduate student at Columbia, Streitwieser became familiar with Michael Dewar's 1945 paper in *Nature* on the structure of stipitatic acid. This was a watershed in his career. Stipitatic acid contains the then-novel tropolone substructure. As Streitwieser recalls, "Dewar's paper was important in my development for showing the power of theory to produce useful chemical insight in a time when organic chemistry was just emerging from a wholly classical period."

Streitwieser's own research focused initially on physical organic problems but soon evolved into a blend of experimental and theoretical organic chemistry. "Writing this book has been valuable to me," he says.

"It pointed out that my scientific life has been pretty positive. I had done something. I just don't sit down and review, evaluate my career like that normally. I think that I would really enjoy reading this book!"

Many organic chemists have pored over his famous texts, "Solvolytic Displacement Reactions at Saturated Carbon", published in 1956, and "Molecular Orbital Theory for Organic Chemists", published in 1961 (which Roald Hoffmann once described as "one of Streitwieser's most important contributions to organic chemistry, for it alerted many organic chemists to the power of theory"), not to mention his undergraduate textbook co-authored with his Berkeley colleague Clayton Heathcock and later also with Ed Kosower.

"Writing a book with someone is probably the most intimate thing, next to being his wife," joked Heathcock. "Andy has an incredible instinct, a deep understanding of chemical principles. He can see through to the core of a problem, strip through the superfluous camouflage. He has a deep curiosity about many things, and he is interested in almost anything you might want to talk about."

His former student, Stanford professor John Brauman, pondered, "What really stands out about Streitwieser? Aside from his scientific acumen, about which there is no debate, he is an extraordinarily good mentor. He allows his students to develop on their own. He treats people with respect, and he is interested both in their research and in their education. I have tried to behave toward my own students the way Andy does to his students."

For the past 40 years, Streitwieser has encouraged many of his graduate students to pursue research in both the experimental and theoretical domain. His work on uranocene is perhaps the crowning example, for the molecule was "conceived in theory" and then prepared on the bench in his laboratory! Streitwieser's good friend Paul Schleyer said recently, "If you want to compete, compute." One might argue that a complete chemist must be both an experimentalist and a theoretician. Streitwieser and his students have led the way in rare form.

"Science has changed over the past 40 years," notes Streitwieser. "Experimentation is instrument-controlled. Today, being a professor is more businesslike; one must be an accountant, having greater financial accountability. Now, if you can't get a grant as a young professor, it really counts against you. Young people are under more time pressure, they must work harder, nights and weekends. I didn't think I worked that hard as a young professor, but in retrospect I really must have!"

Clearly, there are many tasks undertaken by chemists that appear to benefit the profession more than the individual—and which, in fact, may detract both from the individual's research productivity and consequently from scientific progress. As Streitwieser explains, "I did not include my editorial activities in *Progress of Physical Organic Chemistry* in this autobiography. This had been a time-consuming chore, and I don't find it particularly interesting nor did it have any significant positive impact on my research." It is sometimes difficult to evaluate trade-offs between individual research productivity and donations of time for the general good. In addition, as Streitwieser reflects, "One cannot be doing something significant all of the time! It is easier to do seminal work in the atmosphere of a great school, of an MIT or a Caltech or a Berkeley. Being around people thinking big thoughts all the time, it rubs off. Stimulus. Self-generation is so much harder."

When asked to name a few scientists he has most admired, Streitwieser responded, "Glenn Seaborg. He is an excellent administrator, is introspective, analyzes the world. He kept a journal; I wish that I had done that. My old and dear friend, John Prausnitz, a widely read, excellent chemist. Jack Roberts combines excellent chemistry with being a good human being, friendly, forthright. The late George Pimentel, a mensch, natural, friendly, responsible."

Streitwieser himself is a man of many interests. He has what he describes as a "serious" interest in photography, and selections from his award-winning collections are included in this volume. He is not a construction worker, yet during his sabbatical year in 1959–1960, he remained in Berkeley and partitioned his time between writing "Molecular Orbital Theory for Organic Chemists" and completing a bedroom, playroom, and bathroom in the downstairs unfinished area of his new home. "This combination of some physical activity juxtaposed with writing is well-recommended as a way of dealing with a large intellectual project."

He says he is not an athlete. But Susan protests, remembering how frequently she played racquetball and David played ping pong with him. And of the pictures he has supplied for inclusion with this book, one is of himself playing football at a group party and a second is of him whitewater rafting.

Howard Mel, a close friend for many years, says, "He has great equanimity, a calm and reasoned approach. He is both straightforward and positive. I admire his remarkable organizational skills. Another side to Andy and Sue is their devotion to opera, as epitomized by their frequent travels all over the world, seeking out many performances of Wagner's 'Ring Cycle.'"

Streitwieser has a keen sense of humor. I asked Sue what it is that makes Andy "Andy". Laughing, she told me of a party he went to with his first wife. "All the men stood barefoot behind a sheet. Only their feet were showing. It was great fun. I certainly would recognize his feet anywhere! Seriously, he is a wonderful mix of many things: chemist, outdoors man, philosopher. He is a good citizen with a clear conscience."

Streitwieser's daughter reveals another side of him. "Not many people know about Dad's love of sheer volume. I used to play in a very loud punk/pop band, and Dad would often come see us at the local club. Years later, we were discussing the band, and the song he named as his favorite was definitely the loudest, most dissonant song of the bunch! While both Mom and Dad appreciate my more tasteful folk-influenced current band, I think Dad misses my louder, wilder music. I've also seen him get completely worked up at a Taiko concert, and when Mom leaves the house, I know he loves to crank up the stereo and soak in the volume like a teenager."

"The creative moments, these have pleased me the most in my career," Streitwieser muses. "I recall driving with my wife Sue after the uranocene work; the feeling was terrific. Being with Sue and being able to communicate with her added to that pleasure. Many years earlier, when I was almost finished with the solvolysis review, I recognized that I had created something really good. 'On the map,' I thought. With uranocene, being on the map wasn't so important. It is the ecstasy of creativity!"

"When is Andy overwhelmed with happiness?" Sue, his wife mused. "When catching a 20-inch trout, that's Snake River heaven! And when the chemistry really works, like the uranocene."

His son, David, agrees: "Dad relishes, enjoys everything he does. Fly-fishing in Idaho is a passion, a science for him. He matches insects and techniques to the constantly changing situation. It is intellectual as well as recreational. He is always thinking, planning. In whatever Dad does, he is an avid partaker, enjoying, getting the most out of it. He is always looking to enrich his life."

And these words from Streitwieser himself are perhaps the best summary of his attitude and his zest for this world. "If I had one wish, it would be to stay healthy for many more years, because I enjoy living so much!"

Long life to you, Andy!

JEFFREY I. SEEMAN
Philip Morris Research Center
Richmond, VA 23234

October 18, 1996

*A Lifetime of Synergy
with Theory and Experiment*

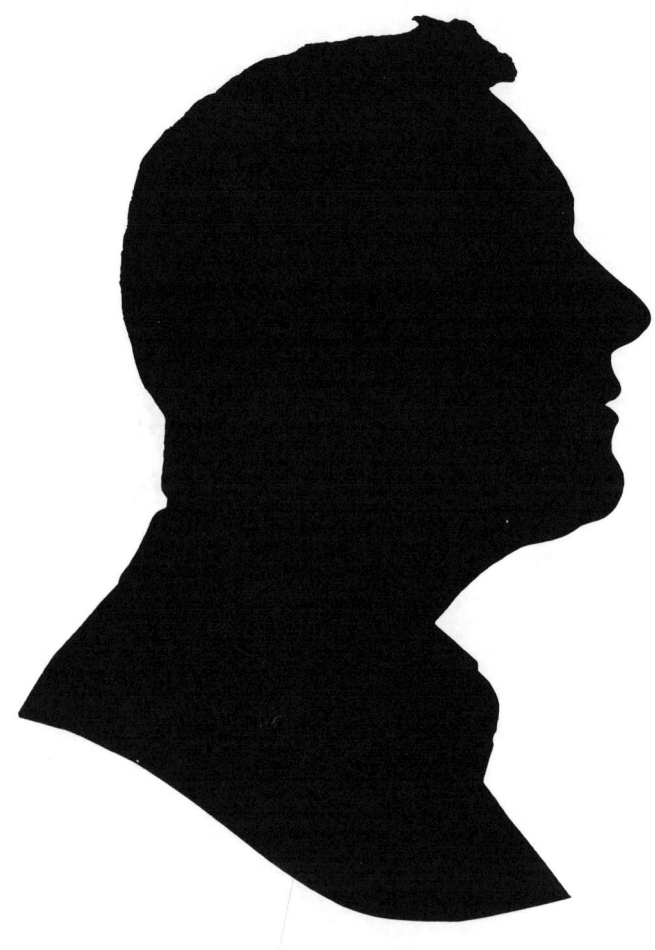

# *About the Writing of This Book*

It's hard to believe that this project is finally approaching completion after its first inception almost a decade ago. At the beginning of Jeff Seeman's project, we were to be separate short chapters in a single book. My initial writing was organized accordingly. As the project developed it was to be a multivolume work with several chemical autobiographies in each volume, then two per volume, and as the first volumes appeared, each autobiography became a separate volume. My own profile at first took my chemistry only to the mid-1970s but had to be brought up to date so that it now includes ongoing projects.

The idea of an autobiography is rather daunting. One thinks immediately of narcissism and ego trips. Clearly, an element of "Look, Ma, no hands" is involved. But these were to be chemical autobiographies, at least that was the emphasis when I started my own profile almost 10 years ago. With each rewriting Jeff urged me to add more and more: more background, more personal views, more personal biographical material. The original manuscripts were mostly chemistry with little material on my personal life. I was scheduled, as one of the youngest of the list of profilers, to be one of the last published. Meanwhile, others in Jeff's project were published, and I have benefited from the feedback of the early volumes to make additions to my own. One important influence is that of Carl Djerassi's volume. His inclusion of his daughter's suicide has encouraged me to include personal details of my first wife's suicide and its effect on my chemistry. This material had been omitted entirely from the early drafts. I have added this material reluctantly because it is important in understanding my chemical development. It is easy to talk about one's achievements and triumphs, but pain is personal and private.

The primary emphasis, however, is still on chemistry: why I worked on different projects, where the ideas came from, and how events in my personal life affected my chemistry. The title was chosen before "synergy" became such a clichéd word, but it still seems appropriate. The *American Heritage Dictionary* defines "synergy" as:

1. The interaction of two or more agents or forces so that their combined effect is greater than the sum of their individual effects.

2. Cooperative interaction among groups, especially among the acquired subsidiaries or merged parts of a corporation, that creates an enhanced combined effect [from Greek *sunergia*, cooperation, and from *sunergos*, working together].

Both parts of this definition fit my chemical life, a life in which theory and experiment have interacted so that both were enhanced. The "theory" involved was not always molecular orbital theory; it has often been nothing more elaborate than simple applications of Coulomb's law. In fact, I believe that one of the continuing themes has been a renewed appreciation of simple ionic interactions in chemistry. The second part of the definition is also appropriate. Many of my students have done only experimental chemistry, some have done only theory, many have done both, but all have been exposed to theoretical interpretations and experimental detail. I think that this aspect of "synergy" has enhanced their own chemical understanding.

As one of the last profilers, this account has been a long time in the making. I have in the meantime retired (more properly, I became "emeritus"; I am not really retired). My research group has shrunk, and some projects are completed or in their final stages but not yet completely written up and published. There's still a lot of backlog to go through. Accordingly, most (but hopefully not all by any means) of the tales in this volume are finished stories. I hope that they will contribute to the history of modern organic chemistry and indicate some of the trends, as I see them, of the future.

# A Lifetime of Synergy with Theory and Experiment

*Our intrepid editor, Jeff Seeman, and me after a luncheon in which we discussed philosophies of life as well as finishing this volume. The time was during the ACS meeting in Anaheim, March 1995.*

One important further point needs to be made. More than 200 undergraduates, graduate students, postdoctorates, and visiting professors have worked with me in the course of my career. I could not mention all of them by name and, in fact, could not include all of their chemistry. Many individuals, however, have been mentioned. Some have interesting personal stories and others add a personal note to otherwise cold chemical facts. My story is the story of many people, including many who are anonymous in this account.

Writing this book was assisted by grants from the National Science Foundation and the National Institutes of Health. I am indebted to several readers of drafts whose helpful comments have been incorporated in the final product: several of my students; Professors Edward Kosower and William Dauben; Professor Clayton Heathcock and his wife, Cheri Hadley; my wife Sue, daughter Susan, and brother Bill; and, of course, our hard-working and always prodding editor, Jeff Seeman.

# Before College

## Early Years and Organic Specialties

Quite a number of us at Stuyvesant High School, New York City, in the early 1940s had learned a reasonable amount of organic chemistry entirely on our own. We used to play games in which we would construct complex acyclic carbon skeletons with various functional groups and then try to assign names in accordance with IUPAC rules. We knew these rules from the *Handbook of Chemistry and Physics*. This book even then was a great bargain, and most of us had a copy. Most of us also had home laboratories of varying degrees of unsophistication and had accomplished several organic preparations. Thus it was that we could see how many compounds on Armour's "wanted" list could be readily synthesized. Each week in *Chemical & Engineering News* the Armour Research Foundation published a list of "wanted chemicals". Few specialty chemicals were available commercially in those days; Aldrich, Columbia, Farchan, etc., didn't exist.

Two enterprising colleagues of mine, Lester Friedman° and Edward M. Kosower°, perceived this need and suggested a three-way partnership, to be known as "Organic Specialties", to fill it. I resisted this venture into industrial chemistry until they showed

---

° *See* Appendix A for brief biographical information about each person designated with the ° symbol.

*Ed Kosower (left) is one of my oldest friends. Here we are carrying out an experiment in Stuyvesant High School, around 1944.*

me a firm order from the General Electric Company for 100 g of 1,3,5-trichlorobenzene. They had used Organic Specialties letterhead stationery that they had designed and had printed and with this stationery had responded to an Armour Research Foundation ad. We were all aware that symmetrical trichlorobenzene was readily prepared by chlorination of aniline followed by deamination and thought that filling such an order would be straightforward.

I invested my $11, and the three of us set to work to fill the order. We bought a lecture bottle of chlorine and suitable fittings, and we carried out the chlorination in an open lot across from my home in Jamaica, Long Island. In those days Queens County had large tracts of empty land. The large empty lot across the street from my home had some foundations for homes unbuilt that had been there from time immemorial. The concrete platforms in the open air served as an excellent manufacturing facility for the chlorination.

We three each had home laboratories. Ed's was the most limited because he lived in an apartment in Brooklyn. Lester's was the most extensive and occupied a portion of the basement of his parents' lingerie store in Manhattan. I always felt a little strange walking straight into this store, past the ladies shopping for intimate feminine apparel and to the back door that led downstairs to the more comfortable and familiar environment of Lester's laboratory. This shop was close to our school, and it was a convenient place to meet and to discuss chemistry despite the gauntlet required.

My laboratory was in the basement of our house. This laboratory had started with a Gilbert Chemistry Set in 1938 when I was 11 years old. The experimental work possible even with this set lured me away from my first scientific love, astronomy. Science for me started when I was 9 years old. I had seen horoscopes in the newspaper and wanted to learn more about astrology. I took out a book from the public library that taught me about stars and constellations and the legends on which they are based. This book was an excellent one for children, and I'm sorry I can't recall the title or author. It also presented the elements of astronomy and taught me the difference between it and astrology. This book provided me with my first direction toward science.

I started going to the Hayden Planetarium and joined the Junior Astronomy Club at the Museum of Natural History in New York. I thought nothing of traveling from the Bronx, where we lived until we moved to Queens the following year, to Manhattan by subway when I was 10 years old, even for meetings at night, a prospect that I would find more daunting now a half-century later! Berkeley students know that I am not bashful about asking questions in seminars. This trait was manifest even at age 10. I recall one astronomy club meeting at which G. Edward Pendray° talked about the early rocket experiments of the American Rocket

Society. He also talked about cosmic rays, and I ventured to ask if these particles damaged our molecules and cells as they passed through us. Unlike my seminar questions nowadays, however, on that occasion I was so nervous about getting up to ask the question I didn't really hear the answer. We shall see later that Mr. Pendray had an important although indirect effect on my subsequent career.

My early venture into astronomy gave me my first experience in public speaking. My fifth-grade class in P.S. 85 in the Bronx was unique. It was a natural history classroom with a menagerie of guinea pigs, a rabbit, and even a one-foot baby crocodile that we had to force-feed by hand. Our teacher, Ms. O. M. Decker, knew Dr. Roy Chapman Andrews, and he had given her a dinosaur bone that was displayed on one wall. That class certainly helped to confirm my interest in science. One of our fifth grade activities was a weekly event in which we sat in an auditorium and listened to a citywide radio series on science and natural history. One such program was devoted to the stars, and Ms. Decker encouraged me to prepare some introductory material in a talk that I presented to several classes as a preparation for the radio broadcast.

My interest in astronomy continues, and I still enjoy the sight of familiar planets and constellations. Stargazing, however, was no match for the colors and transformations I could accomplish with chemicals. This type of fun with chemistry can no longer be had by children to the same extent as was possible in my youth. We trust our children to make the right decisions in a world filled with the temptations of drugs and other evils; yet all but the most innocuous chemicals are considered too dangerous (or perhaps too subject to lawsuits) to form part of their play and learning.

I had also received a microscope set together with the chemistry set that Christmas of 1938. I was fascinated to look at minute structures through the microscope. The combination of the stars at night, microscopic structures, and colorful transformations with chemicals was an ideal introduction to science. I'm sure the same prescription would work with children today.

That first small chemistry set was augmented by a larger set for my birthday the following summer. Thereafter, I acquired chemicals and apparatus from shops in Manhattan and built a working facility. My parents were not happy about the smells that were created from time to time (hoods were unheard of in home laboratories!). My father was a carpenter who was born in Munich in 1901. My mother was born in the small village of Obertal in the Black Forest in 1904. Both immigrated to the United States in 1923 during a time of revolution and inflation in Germany. My mother was only 19 at the time. My parents met and married in Buffalo, NY, where I, in 1927, and three years later, my brother, William, were born. With the developing Depression in 1930, my father lost his job at the Pierce-Arrow automobile factory and moved us all to New York City in hope of better employment possibilities. But employment was spotty, and for a number of years we were quite poor. In addition to his carpentry work, when he could find it, my father was the superintendent of an apartment building. This job at least paid the rent but little more. There were times that I know my father carried his tool box for many blocks just to save the five-cent subway fare.

The New York World's Fair of 1939–1940 proved to be a turning point. My father found steady employment in construction in the fair and was able to save the $100 required for the down payment on the house in Jamaica. Somewhat later he took the opportunity to run the local franchise of the Roto-Rooter Company. He learned the sewer cleaning business and became a small-business manager. This change provided some valuable experience for me as well.

Stuyvesant High School ran two sessions a day, and during my second year I was able to go to the morning session. Taking morning classes required getting up early in order to make the one-hour commute and start school by 8 A.M., but it gave me free afternoons and, like many of my colleagues, I was able to find a part-time job working in the city. My job with the Buttoncraft Company involved doing a variety of odd jobs but mostly packaging buttons and delivering them to garment industry establishments in Manhattan. Buttons by the thousand can be quite heavy, but I became adept at hauling a handcart through the city streets. My salary was 50 cents an hour. Much of it I saved for college. This job lasted for a year, and then I was able to quit and work for my father. Few of my colleagues know and some will

*My father and me, late 1927 or early 1928. My father died in 1967 of a stroke after an earlier stroke in 1959. He had been a heavy smoker.*

undoubtedly be amused to learn that I cleaned many a sewer in my young days. In fact, even after I had earned my Ph.D., on a visit home I went out on a few jobs to show that I had not lost my old skills. The knowledge of plumbing that I gained with that experience was useful years later when I did some remodeling while writing my first book, but that story comes later.

I was named after my father and am therefore Andrew Streitwieser, Jr., with no middle initial. This practice of naming sons after fathers leads to confusion, and I don't recommend it. Persistent attempts have been made, undoubtedly because of the Junior, to cite me with a middle initial of "J". There are few Streitwiesers in the world, and having an unusual name is an advantage. So far as I am aware, my son David (who is now a practicing physician) and I are the only Streitwiesers in the scientific literature. David was a biochemistry major at Berkeley, and his undergraduate research with Professor Bruce Ames resulted in a publication. In recent years I have started to delete the "Jr." from my name because my father died more than 25 years ago.

In English-speaking countries we tend to "mispronounce" our name as "Strytwyser", but in Germany I use the correct Germanic pronunciation of "Strytveeser". The name literally means a disputed field and is derived from *Streit*, a fight or quarrel, and *Wiesen*, fields or meadows. In a valley near the Danube river and the city of Melk, Austria, there is a small ruin of

# A Lifetime of Synergy with Theory and Experiment

With my roadster in Buffalo, 1930.

My mother took me to Germany in 1930 when I was 3 years old. It was her first trip back to Obertal since she left in the early 1920s. She was pregnant with my brother Bill at the time of this picture.

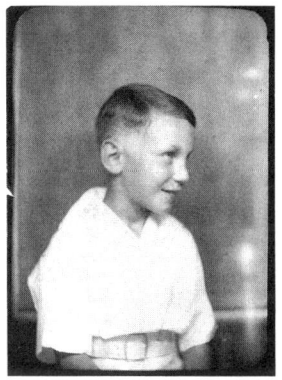

A school picture of me at the age of 6.

a castle, Burg Streitwiesen, above the hamlet of Streitwieser. The castle is now a youth hostel, but it was an active castle during the 12th to 14th centuries. According to legend two brothers, one good and one evil, fought, and the bad one won. The castle degenerated and the Streitwiesers scattered. Talk about family skeletons! I understand that a number of Streitwiesers live in Laufen, a suburb of Salzburg just over the German border. My

*My parents, about 1946. At this time my father ran a franchise of the Roto-Rooter Company on Long Island. Our Jamaica house is in the background. When I visited this house again years later, it seemed very small.*

# A Lifetime of Synergy with Theory and Experiment

*My mother, Sophie, in 1988, at the age of 84. She stayed with us in Idaho that summer before returning to her Florida home where she lived by herself. She suffered a stroke the following year and could no longer live by herself. She lived in a retirement facility in Berkeley where we visited her regularly until her death in November, 1995.*

paternal grandfather was born in Laufen but was raised in Munich. My mother's family can be traced back for many generations as farmers, hunters, and town officials in a small part of Germany near Freudenstadt.

My parents' success in overcoming the deprivations and hardship of their first years in the United States is a tribute to their innate abilities and courage, and I am proud of their accomplishments. Their understanding, implied if not realized or expressed, also showed in their forbearance of my scientific activities. Neither of my parents had any concept of science or of what I was doing or planning. Both had barely finished grade school with the disruptions of the first World War. Nevertheless, they indulged my interests and helped where they could. My father installed a gas line so I could use a Bunsen burner in my home laboratory. I did not have running water and had to use two pails and a siphon arrangement with my condenser in reflux and distillation procedures. My balance was homemade and crude; my weights were coins. Nevertheless, a number of chemical experiments and preparations were accomplished in this laboratory. And I taught myself to keep a notebook.

I still have my notebooks from 1943 and 1944. I met Ed Kosower in late 1942 when he transferred to Stuyvesant from junior

*In 1980, I visited Burg Streitwiesen, a castle above the hamlet of Streitwieser in Austria. The small castle is now a youth hostel run by the Austrian government.*

# A Lifetime of Synergy with Theory and Experiment

*Edward Kosower at MIT in 1946 with the President of MIT, Karl T. Compton, looking on. After graduating from MIT in 1948, Ed Kosower received his Ph.D. at UCLA with Saul Winstein in 1952 and did postdoctoral research with Cyril Grob in Basel, 1952–1953. He was on the faculty of Lehigh University, the University of Wisconsin, and SUNY, Stony Brook, before moving to his present position as Professor of Chemistry, Tel-Aviv University.*

high school. I sometimes went to his house in Brooklyn, but more often he came to Jamaica and we did experiments together that are dutifully recorded in these notebooks. A typical entry is that for January 1, 1944: "Eddie came over and we made some fluorenone." In most cases no reason was given for the experiments. Although some of our organic preparations were quite sophisticated, we clearly lacked direction. Experiments should have an objective. The Organic Specialties experiments did not share this limitation; here the objective was purely commercial.

For the General Electric order for trichlorobenzene, the chlorination of aniline in our "outdoor facilities" was not without incident. The chlorination produces HCl that reacts with excess aniline to give the solid aniline hydrochloride, which repeatedly clogged the delivery tube. We had to try to stir the mass by swirling and to unclog the tube while chlorine flowed out, but we finally got it done. I later worked up the product and passed it on to the next person for the deamination. One thing we learned from our experiments and particularly in this venture is that organic chemistry in the laboratory is sometimes quite different from organic chemistry on the printed page. The preparation did not go as smoothly as we had planned, and we ended up with only 50 g of yellow material, for which we did receive part payment. This disappointing result taught me that there was much still to be learned about laboratory work. I therefore left this foray into industrial chemistry and retired to my laboratory to do more academic-type research.

One other incident points up further the innocence of the age. We had located a reference that we needed in a Russian journal. This journal was not, of course, available in our high school library, nor was it to be found in any of the branch libraries of the New York Public Library system. It was available only in the main library at 5th Avenue and 42nd Street. High school students, however, were not allowed to use this library. We got around this restriction by carefully typing a letter on Organic Specialties stationery requesting that the bearer be allowed to use the main library for company purposes. The letter was signed by Lester Friedman as President. The ploy worked, and I was able to spend a number of hours in the main library on this translation that I did virtually letter by letter with the aid of a Russian dictionary.

My interest in theory was apparent even at the age of 17. The New York school system had little in the way of counseling in the 1940s, and our activities in organic chemistry were totally independent of school and teachers. Nevertheless, by exchanging information among ourselves and with the public library system, we managed to learn quite a bit of chemistry beyond the high school level. I remember reading A. E. Remick's *Electronic Interpretations of Organic Chemistry*.[1] I don't recall how I learned of this book, but I bought my own copy while in high school (and I still have it). It's a wartime edition with thin paper and narrow margins. Although much of this book was beyond me, I was fascinated by the applications of resonance structures, especially in aromatic chemistry. Aromatic chemistry was popular at that time, particularly for us beginners. Many of the preparations were straightforward, and most of the derivatives, particularly for polycyclic benzenoid compounds, are crystalline and did not require elaborate apparatus. Vacuum distillation, for example, was a technique not generally available in our home laboratories. Moreover, many of the starting materials and reagents were available as free samples from various companies (especially when solicited with Organic Specialties stationery!).

## My First Paper

I had acquired a sample of fluorene from the Koppers Company and became interested in its chemistry. Indeed, I did an extensive literature search of fluorene chemistry using *Chemical Abstracts*. Most of this chemistry I did not understand, and little did I realize at that time the important role that fluorene derivatives would serve as indicators in my subsequent research on carbon acidity. But in my survey I came across an early research paper and a later review article[2,3] by Herbert C. Brown° on chlorinations with sulfuryl chloride, a substantial amount of which I had recently acquired as a free sample from the Hooker Company. Professor Brown had reported that the reaction was applied to fluorene but that the reaction product was not identified.

Here was my opportunity for some original research because I had all of the necessary materials at hand. I ran the chlorination and obtained a white crystalline material having a melting point that agreed with that reported for 2-chlorofluorene. I realized that this would not be sufficient characterization for a publication so I made authentic 2-chlorofluorene using the Sandmeyer reaction with cuprous chloride and 2-aminofluorene that I had made previously following the procedures in *Organic Syntheses* (Scheme I). The mixed melting point confirmed the identity, and I wrote the work up as a Note for the *Journal of the American Chemical Society*[4] (Figure 1). I recall the thrill in sending my first completed

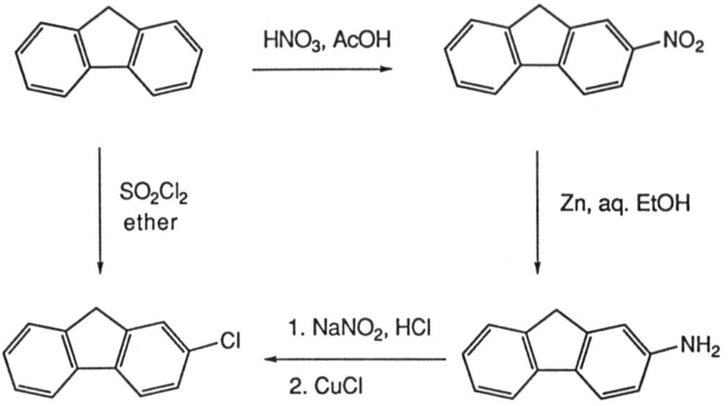

Scheme I.

scientific manuscript to the editor, an exciting and uniquely thrilling feeling of accomplishment. Half a century and more than 300 research papers later, I still get that same thrilling feeling in sending a new manuscript to a journal. Of course, seeing that first paper in print was an indescribable pleasure, but another new feeling came a few months later when I found my article summarized in *Chemical Abstracts*. It was a special thrill to recognize that I was listed in the chemical archives and had thereby achieved a kind of immortality.

In reviewing this work for the present history, I was amused to find in my notebook that the mixed melting point was taken on October 10, 1944, and that the Note was published in the December issue of *JACS* in the same year! Never again would I see such rapid publication of experimental results! This story has an amusing epilogue. After the appearance of my Note, I received a letter from the Armour Research Foundation saying that they had a client who needed some 2-chlorofluorene and asking if I could provide it. After a quick phone call to my former partners, I replied to Armour with the information that I could not provide the material but that I could recommend Organic Specialties as a source. In due course, Organic Specialties received an order, which they then filled. For my services as a consultant I got back my original $11 investment.

There have apparently been some problems in reproducing this sulfuryl chloride reaction with fluorene.[5] I have long hoped to take a personal and perhaps nostalgic look back at this reaction. I

### Chlorination of Fluorene with Sulfuryl Chloride

BY ANDREW STREITWIESER

Sulfuryl chloride is being used increasingly as a chlorinating agent because of the ease with which it reacts and the high yields obtained. In 1939, Kharasch and Brown[1] treated fluorene with this reagent and reported that the fluorene was chlorinated in the nucleus. However, they did not report the position taken by the entering chlorine atom, nor did they give any yields or procedures. The purpose of this work was to determine the position of the entering chlorine and to find out whether this method is a convenient one for preparing the chloro compound.

The fluorene was chlorinated in ether solution with a slight excess of sulfuryl chloride. The product melted at 95–96°.[2] This compares favorably with the 2-chlorofluorene prepared simultaneously by Chanussot[3] and Courtot and Vignati,[4] m. p. 96–97°. Authentic 2-chlorofluorene was prepared by the method of Chanussot[3] from 2-aminofluorene by the Sandmeyer reaction. A mixed melting point determination showed the two to be identical. The following procedure is recommended as a quick, convenient method for preparing 2-chlorofluorene.

#### Experimental

2-Chlorofluorene from $SO_2Cl_2$.—Sixteen grams of fluorene and 120 cc. of anhydrous ether was placed in a distilling flask. Not all of the fluorene dissolved. The flask was stoppered with a rubber stopper carrying a separatory funnel containing 8 cc. (13.4 g.) sulfuryl chloride. The latter was added rapidly to the solution. When all was added, the ether was distilled off, and the residue recrystallized from alcohol. It can also be purified by steam distillation: yield, 85% of a white crystalline powder, m. p. 95–96°.

The author wishes to acknowledge the assistance of Mr. L. Friedman and Mr. E. Kosower in this work.

---

(1) Kharasch and Brown, THIS JOURNAL, **61**, 2149 (1939).
(2) All melting points corrected.
(3) Chanussot, *Añales asocn. quim. Argentina*, **15**, 216 (1927).
(4) Courtot and Vignati, *Compt. rend.*, **184**, 1479 (1927).

80–29 169TH STREET
JAMAICA 3, N. Y.          RECEIVED OCTOBER 13, 1944

*Figure 1. My first publication. (Reproduced from reference 4. Copyright 1944 American Chemical Society.)*

suspect that the purity of solvent and reagents may be important and the chemicals in my home laboratory may well have harbored impurities that served as catalysts or promoters.

## Science Talent Search

The work I did with fluorene also served as the basis for the original project portion of my entry in the Fourth Westinghouse Science Talent Search in the fall of 1944. That year was the first in which Stuyvesant High School competed, and two of us became "trip winners". Ed Kosower and I joined 38 other winners in a trip to Washington, D.C., in March, 1945. I recall that trip as a memorable and extremely enjoyable experience. We met Mrs. Franklin Roosevelt and then Vice-President Harry Truman. Almost as thrilling for the chemists in the group was the opportunity to meet and talk with Roger Adams°. Ed Kosower took the opportunity to show Dr. Adams his proposed synthesis of quinine and was thrilled to learn that it had a good chance of succeeding except for some possible problems with stereochemistry. Ed's father was a New York taxicab driver who was very proud of his son and who told many of his passengers about his son's contact with the famous Dr. Roger Adams of the University of Illinois. It was inevitable that one day, of course, the passenger to whom he told this story was Roger Adams himself.

These forty trip winners were a highly competitive group, but despite the competition for a range of scholarship awards there was an atmosphere of friendship and good fellowship within the group. My diary of the trip reveals my own competitiveness, but it was great fun being with such bright and accomplished people. We went sightseeing together in Washington and listened to special lectures by such famous scientists as (the late) Harold Edgerton and Harlow Shapley. I remember vividly my feelings at the Awards Banquet when the awards were being announced, starting with the eight intermediate awardees. As each name was announced, I had a mixed feeling of anticipation about whether my name would be next. On the one hand, it would be satisfying to receive an intermediate scholarship, but on the other, it would mean that I would not be the top awardee. In

# A Lifetime of Synergy with Theory and Experiment

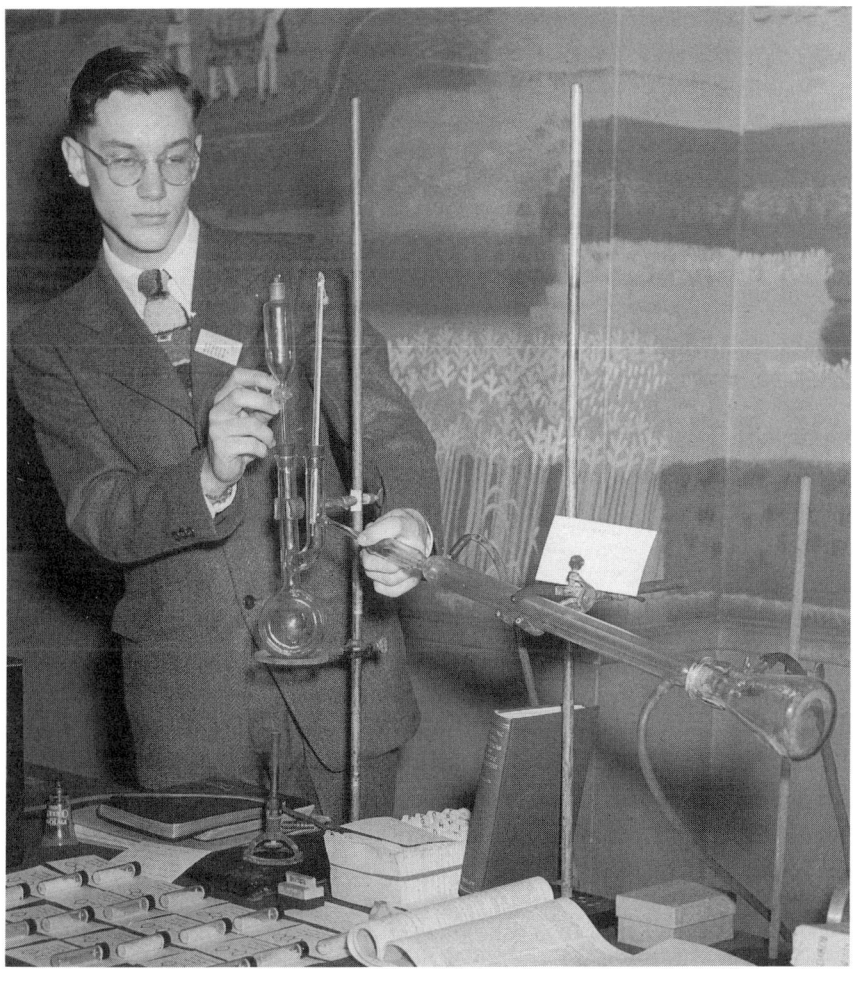

*One of the Westinghouse Science Talent Search activities is a demonstration of our science projects. I am setting up my project on my experiments with fluorene in Washington, DC, March, 1945. (Reproduced with permission from Science Service, Inc.)*

the event, my name was called as an intermediate awardee, and I was thrilled. Moreover, Ed Kosower received the top scholarship award. Thus, Stuyvesant High School did well in its first outing with the Science Talent Search. This experience was my first with a group of national peers, and the event was important to my career by showing that I could "belong" in such a group.

The Science Talent Search was developed by Watson Davis

of Science Service together with Westinghouse's G. Edward Pendray, the same Pendray whose lecture about the experiments of the American Rocket Society I had heard in my junior astronomy club years before. The Science Talent Search recently celebrated its 50th year. Many thousands of students have been encouraged and rewarded by this institution; a large number have gone on to the Ph.D. degree, and several trip winners are now Nobel Laureates. Some other organic chemists who have been Science Talent Search finalists include Professors Ronald Breslow (1948), Columbia University; Herbert House (1947), Georgia Tech; Andrew Kende (1948), University of Rochester; Lawrence Schaad (1948), Indiana University; Charles Wilcox (1948), Cornell University; and Dr. Edel Wasserman (1950), DuPont. I am impressed by the scientific sophistication of the projects contributed by recent winners, many of whom have had the opportunity to work with modern equipment in collaboration with successful scientists. During these years Stuyvesant High School has been one of the top producers of finalists among high schools, with 70 through 1990. Ed Kosower and I could not have appreciated at the time what a successful tradition we had initiated at Stuyvesant, but it does provide a satisfying feeling today.

Other now-famous chemists were in Stuyvesant High School at that time. Robert Zwanzig° and Benjamin Widom° were contemporaries, and Jerrold Meinwald° was two years ahead of us. Martin Saunders° was three years behind us. Stuyvesant High School is one of several citywide specialty high schools in New York City. Others specializing in science are the Bronx High School of Science and Brooklyn Technical High School. Teachers in my grade school in Jamaica, P.S. 131, recommended Stuyvesant to me, and I took the entrance examination required for admission. Stuyvesant at that time was for boys only, but it became coeducational in recent years. In 1992, the school moved from its original location at 15th Street and 2nd Avenue to new facilities just north of Battery Park City in Manhattan. I am sure that the new school will be as successful as the old in providing encouragement and training for new generations of imaginative and productive scientists.

# *Higher Education and the Army*

## Columbia College and Army Service, 1945–1948

During my last year in high school I applied to and was accepted by Columbia College. My parents understood the value of formal education despite their own lack of it. They had hoped that I would be able to go to college, but for many years this hope had a low probability of fulfillment. The idea of postgraduate studies simply did not exist in our world of that time. My poor brother, of course, suffered by my relative prominence. No thought was ever given to the possibility of his going to college. I suspect that my leaving for the Army in 1946 was a blessing for him and allowed him to take "center stage". In the event, we both graduated from Columbia College. He subsequently obtained a Master's degree and became a highly successful high school teacher in Long Island specializing in history and current events.

    I had chosen Columbia College in order to save money by living at home. The amount I had saved from my employment at Buttoncraft and Roto-Rooter was enough to pay for my initial years at Columbia. I did not realize at the time how important it was for me to get the type of liberal arts education that Columbia provides. The Contemporary Civilization and Humanities sequences required for the A.B. degree opened my eyes to new worlds. I remember my first day in Humanities with Professor Weaver, a dynamic speaker and marvelous lecturer who loved his subject. In our first lecture he presented for discussion the

*A recent picture of my brother Bill (left) and me.*

personality of Helen of Troy. The concept that a fictional person could be discussed in the same terms as a real person, with a personality, memories, and fears, was, despite my having read many novels, a new thought that I hadn't really considered before. My initial skepticism gave way to fuller understanding. Professor Weaver was the dominating nonchemistry influence from my first year in college. The reading in the Contemporary Civilization course of original documents important in Western civilization also added depth to my knowledge of history and of government institutions. This course, especially after I returned from the Army, was another major influence. I regret that such courses are no longer required at many so-called liberal arts colleges.

My undergraduate education at Columbia College was interrupted by a year and a half in the U.S. Army Medical Corps, 1946–1947. Although I became 18 years old in June 1945 and was thus subject to the draft while the war in the Pacific was still in progress, I was deferred to the end of the fall semester and was finally drafted in February, 1946. The war was now technically at an end but was not yet legally over. Thus, I received the benefits of the G.I. Bill when I did return to college.

## A Lifetime of Synergy with Theory and Experiment

In retrospect, my Army career was an important experience for me. For the first time I was forced to live away from home and to be more self-reliant. I had never been much of an athlete, but I was in sufficient physical shape to take pride in being able to succeed in the rigors of basic training. I was even awarded a sharpshooters' medal for my prowess at the target range. The Army then sent me to Medical Technician's training, after a variety of tests determined that my military aptitude lay in that direction. By the time a trainload of us arrived at Fitzsimmons General Hospital just outside Denver, CO, all of the various

*In the U.S Army Medical Corps, 1947. I was drafted in February, 1946, and spent 1½ years in the U.S. Army Medical Corps. For most of that period I was stationed at Fitzsimmons General Hospital, Denver, CO, where this photo was taken.*

training schools had closed because the demand had evaporated with the ending of the war. Because there was nothing for us to do, we were placed in a holding company while the Army decided what to do with us.

Most of my contingent ended up as hospital orderlies, but I did something one is warned never to do in the Army: I volunteered for a special rush project requiring typing skills. My colleagues today who see me typing with two fingers might find my assignment as a typist rather incredible, but remember, this was the Army! Many civilian employees were being phased out at this time, and their duties were taken over by soldiers. Thus, I was finally assigned to the Admissions and Dispositions (A&D) office of the hospital with a night duty, the primary responsibility of which was to turn out by 6 A.M. a number of mimeographed copies of the "A&D Sheet", an inventory of all changes in all of the wards. This job had required a staff of four civilians who worked all night. An Army buddy, Paul Renton, who also had had some college experience interrupted by the draft, and I were able to organize the job in such a way that by diligent work one of us could turn out the inventory in four hours, an excellent example of what incentive can do.

This organization gave us a great deal of free time that I used by taking math courses and French at the Denver extension division of the University of Colorado. I also took a subscription to the Denver Symphony Orchestra, went to concerts at the Red Rocks Amphitheater, and did my first skiing at Loveland Pass. Several of us occasionally got together and rented our Sergeant's car to tour the beautiful surrounding country. Denver had no smog in those days, and we could routinely see the Rockies from the eastern suburb of Aurora, a treat rarely possible today.

My Army experience could certainly have been much worse. I was fortunate to be in a beautiful location with time for my own activities. The risk was mostly the continued exposure to patients with active tuberculosis, a common disease through the 1950s. I made sure that this threat was monitored by regular physical examinations for years thereafter. One negative event, however, that still arouses some anger was the denial of my application for student membership in the American Chemical Society on the grounds that because I was in the Army I was not a student! I had hoped to start subscribing to *JACS* but had to wait for my return to Columbia.

On my discharge in August, 1947, I had my first contact with California. On the day of my discharge a patient was scheduled to be flown to California. Through friends I was given a last-minute opportunity to join this flight. I quickly decided to postpone returning home for two weeks and flew in a nearly empty Army DC4 to the Fairfield–Suison Army Air Force base. From there I took a bus to San Francisco and then to Palo Alto, where I arranged to stay for several days in one of the dormitories, again through a friend who was taking a summer session course. On the way the bus passed through Berkeley, but I did not stop to see the University of California campus, never realizing that only a few years hence I would be starting a life-long career there! I was able to do several days of sightseeing in San Francisco, and my terminal leave junket took me by train to Los Angeles and Chicago before returning home in time for the fall semester of 1947.

I completed my undergraduate work at Columbia College and continued with graduate work at Columbia University. I think it is normally better to continue one's graduate education at a different institution from one's undergraduate work, but those times during the period around World War II were not normal. I had graduated from high school in 3½ years by going to summer school and started at Columbia College in early 1945. The plan was to get in as much college as possible before being drafted, and taking 18–20 units a semester was not at all unusual. As a result, I'm afraid that I don't have a great deal of sympathy for students now who complain that 15 units is too much work. The plan worked out well in my case; by the time I was drafted, I had completed two semesters and a summer session and was almost a junior. I also received additional credit for the courses I took in the Army and was able to graduate at the end of the summer session of 1948. That is, I finished college in 3½ years, including the 1½ years in the Army, but my college education was fragmented and my science training had some gaps.

I don't recommend this type of education. Moreover, this fragmentation and the long commute to Jamaica greatly limited my campus life. There was a recreation hall at which I learned to

*Ben Widom and I were seated next to each other at the awards dinner in Las Vegas in 1982. Ben (right) is a theoretical chemist at Cornell University, and we were classmates in Stuyvesant High School as well as undergraduates together at Columbia College.*

shoot some pool. I also learned to play bridge, which provided a helpful respite. In one semester I had laboratory four afternoons a week. The fifth afternoon I relaxed by playing bridge. Nevertheless, my undergraduate life was not as rich in memories as for most students. Several of my friends from that period have become famous chemists. Physical chemistry laboratory in my final year was shared with Benjamin Widom (we were partners together for many experiments. Ben is a theoretician, and he claims that only because I did the experimental work did he complete the course and achieve success as a chemist!), Gert Ehrlich°, and E. Peter Geiduschek°. All three have been elected to the National Academy of Sciences.

---

° *See* Appendix A for brief biographical information about each person designated with the ° symbol.

## Graduate School, 1948–1951

Life in graduate school was better and richer primarily because of the depth of personal relationships forged by close contact with individuals, mostly in the same research group, over a period of time.

In those days at Columbia University, chemistry graduate students spent an additional year of course work before starting research. For me, that additional work was important preparation and compensated in large measure for the gaps during college. In one aspect of that first graduate year I feel I was incredibly lucky. Luck is clearly important in life, and I have had my share of good fortune. I am reminded of Thomas Jefferson's remark, "I am a great believer in luck but it seems the harder I work the more I have of it." And, of course, one must do something with luck when it comes. One of our first-year graduate courses was Organic Seminar, in which each of us was given a paper to review from the current literature. I was assigned a paper on the structure of purpurogallin, **1**.[6] In the course of background reading to appreciate this paper, I also read Michael Dewar's° paper on stipitatic acid.[7] In this paper Dewar reinterpreted published experimental data and proposed a new structure based on qualitative theoretical principles based on resonance structures. His structure, **2**, contains the novel tropolone nucleus, **3**, that is part of the structure proposed for purpurogallin. Dewar's paper was important in my development for showing the power of theory to produce useful chemical insight in a time when organic chemistry was just emerging from a wholly classical period.

I regard Dewar's paper as marking the beginning of an approximately two-decade period of brilliant organic chemistry, which saw a number of fascinating syntheses of cations, anions, and "pseudoaromatic" compounds derived broadly from the Hückel $4n + 2$ rule. As implied by the additional resonance structure shown for tropolone, it may be regarded as a derivative of the cycloheptatrienyl (tropylium) cation, **4**, and this cation was a leading player in the emerging drama. Thus, this seminar assignment brought me into the beginnings of this important part of modern organic chemical history. It led to various research projects in my subsequent career and can probably be said to have culminated for me in the conception and synthesis of uranocene (*see* chapter titled f-Orbital Organometallic Chemistry).

After this first year as a graduate student, I joined the research group of Professor William von Eggers Doering° in 1949. Although I went around and talked to all of the organic professors, the decision had actually been made some years earlier. In the summer of 1945, while I was still a freshman at Columbia College, I audited a graduate course in advanced organic chemistry given by Doering. He allowed me to take all of the exams and gave me an unofficial grade (B) even after he discovered that I was taking first-semester introductory organic chemistry at the same time. His course covered reaction mechanisms and organic theory and I loved it, even though, without physical chemistry, I couldn't understand it all. He was an excellent lecturer and has a charismatic personality, and I was hooked. His research group was an excellent one and at an important time in the recent history of organic chemistry. It included a number of great individuals who have since become important names in today's chemistry. The group included Ken Wiberg°, Jerry Berson°, Al Wolf°, Dick Young°, Herb Meislich°, and Chuck DePuy° (four of whom had been elected to the National Academy of Sciences as of 1994). Most of the group was older than I, and their educations also had been interrupted by World War II. They had an incredible *esprit de corps*, and I learned a great deal from them.

I overlapped one year with Ken Wiberg, who impressed me

# A Lifetime of Synergy with Theory and Experiment

*William von Eggers Doering at Yale. Doering left Columbia in 1952, was at Yale University 1952–67, and then moved to Harvard University, where he is now Emeritus.*

with his remarkable organization. He worked efficiently all week while his sink piled up with dirty flasks and beakers. Saturday he spent cleaning up to be ready for work again the next week. He also loved ice cream. Al Wolf was my teaching assistant during the advanced organic chemistry course I took during my first graduate semester. Al was one of the principal influences of my graduate education; I continued to learn from him my entire time

*Kenneth and Marge Wiberg on the occasion of Ken's Pauling Award in Vancouver, 1992. Ken and I were graduate students in the same laboratory at Columbia, and I participated in his award symposium.*

at Columbia. Al has a soft heart underneath a gruff exterior that is combined with a marvelous sense of humor. Dick Young worked right across from me and had a balanced mature nature that provided me with an outstanding role model. Our laboratory was a large room shared with the small group of Professor Louis Hammett; also it was called "Siberia" because of its physical separation from the rest of Doering's laboratories. Jerry Berson was not part of our laboratory, but we regarded him philosophically as one of us. He was serious and fair-minded and had an impressive personality even then.

# A Lifetime of Synergy with Theory and Experiment

*Al Wolf in 1964. We were in the same laboratory in Doering's group at Columbia. Al spent the fall semester, 1964, at Berkeley substituting for me because of my Miller Research Professorship that year. He was present in Berkeley at the very beginning of the "free speech" era.*

An initiation into a new group usually requires a rite of passage. Mine occurred early. Doering loved Chinese food, and many in his group joined him at a local Chinese restaurant before our weekly Tuesday evening group meetings. Chinese food was relatively new to me, but anyone asking for a fork was regarded with disdain. Thus, I had perforce to rapidly learn to use chopsticks, and the facile use of chopsticks became a useful if unanticipated product of my graduate education. On my first dinner I was invited to try some peppers that were claimed to be quite mild. I could not, of course, refuse. I suspected what would come and was not surprised when, on taking a bite, my mouth and throat were flooded with searing fire. But I stoically kept a cool demeanor and remarked that they were indeed mild. Al Wolf then exclaimed, "They must have changed them!" and took a bite himself. He yelped and quickly dove for the water. Only then did I dive for the water myself, but I was now part of the group.

Just before I joined Doering's group, Harold Zeiss had ac-

*Jerome A. Berson in 1949. Jerry and I overlapped as graduate students in the Doering group at Columbia. He did postdoctoral work with Woodward at Harvard, 1949–1950, and is now at Yale University.*

complished the first resolution of a tertiary alcohol, 3,5-dimethylhexan-3-ol, and had studied the stereochemistry of solvolysis of its hydrogen phthalate ester.[8] Solvolysis chemistry was an active field of research at that time, and studies of rearrangements, stereochemistry, and kinetics involving solvolysis intermediates were important to contemporary physical organic chemistry. My research problem was a continuation of Zeiss's work. I repeated and simplified the resolution and did quantitative studies on the stereochemistry of various solvolysis reactions. Solvolysis in methanol, for example, produced 3-methoxy-3,5-dimethylhexane with 60% net inversion of configuration [60% enantiomeric excess (e.e.) in more modern terminology], and this product showed that the tertiary carbocation intermediate is not free and symmetrical before reacting with solvent. In my further graduate work I showed that a number of nucleophilic solvent additives, such as dioxane or acetonitrile, could intercept the tertiary alkyl cation before reaction with solvent to give more extensive racemization. My graduate research introduced me to a number of aspects of stereochemistry and to the extensive literature and chemistry of nucleophilic displacement reactions; this work would play an important role in my subsequent research career.

I still remember vividly one laboratory tragedy that seemed so important at the time. I had prepared several hundred grams of the tertiary alcohol by a large-scale Grignard reaction and had purified it by a careful fractional distillation when I dropped the flask containing the several weeks of work. I recovered what I could with paper towels and solvent, but I knew my emotional state was not ready for further labwork at that point. I just left the lab, took the subway to Central Park, and spent the next hour on a rowboat in the lake. By the next day, I was ready to purify the recovered product, and I obtained enough to continue with the preparation of the hydrogen phthalate and the subsequent resolution. Sometimes we have to accept being merely human and do something nonproductive in order to be more productive later.

Some of the other problems worked on in Doering's group at that time give an indication of the wide scope of his interests as well as the research technology of the day. Al Wolf, for example, studied the mechanism of formation of $\beta$- and $\gamma$-fenchenes by the acid-catalyzed dehydration of $\beta$-fenchol. He treated the O-deuteriofenchol with potassium deuteriosulfate for a minimum time and separated the mixture of fenchenes using a concentric tube column. This type of fractionating column consists of two concentric tubes separated by a small but precise annular space. The column has high efficiency and low holdup but requires rigorous adiabatic conditions for greatest effectiveness. I recall that Al had his column in a cabinet with lots of insulation and with very slow throughput. A fractional distillation experiment took weeks of effort. But he separated each of the fenchenes and determined their deuterium contents by chemical degradation followed by combustion of the degradation products and deuterium analysis of the combustion water by the falling-drop method. Now, of course, this problem could be done in a fraction of the time by the use of gas chromatography, mass spectroscopy, and deuterium NMR. Nevertheless, Al showed that the formation of the $\beta$- and $\gamma$-

# PROFILES, PATHWAYS, AND DREAMS

## Scheme II

β-Fenchol → α-Fenchene

Cyclofenchene + Cation from 2,6-H rearrangement + β-Fenchene

*Scheme II.*

fenchenes could best be explained by a 2,6-hydrogen rearrangement as in Scheme II.[9] The only modification that we would now add would be to consider a number of the carbocation intermediates as nonclassical.

Dick Young's research involved an early example of asymmetric induction. He showed that the Meerwein–Ponndorf–Verley reduction of a ketone with the aluminum salt of an optically active secondary alcohol gives an excess of one enantiomer of the product alcohol. The results were rationalized on the basis of reduced steric interactions in the favored transition state,

*Richard W. Young in 1950. Dick worked opposite me in the Doering group and taught me a great deal. After his Ph.D. he joined the American Cyanamid Company, where one of his achievements was the invention of Neptazane, the first drug designed for the treatment of glaucoma. Subsequently, he held a series of executive positions at Polaroid Corporation, Houghton Mifflin, and Mentor O and O, Company.*

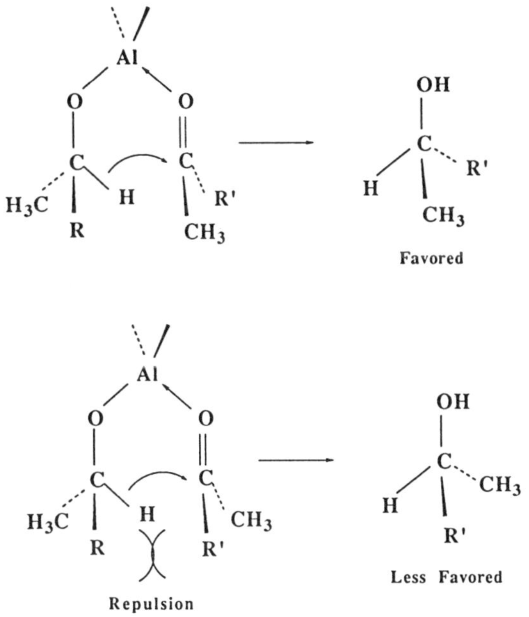

## A Lifetime of Synergy with Theory and Experiment

*Scheme III.*

in which the larger R and R' groups interact with the small methyl groups rather than with each other (Scheme III).[10]

Milton Farber was in the same laboratory with me for my first year, together with Wiberg, Wolf, and Young. Farber's research included the discovery of the reaction in eq 1[11] and exemplifies Doering's interest in the chemistry of bicyclic compounds.

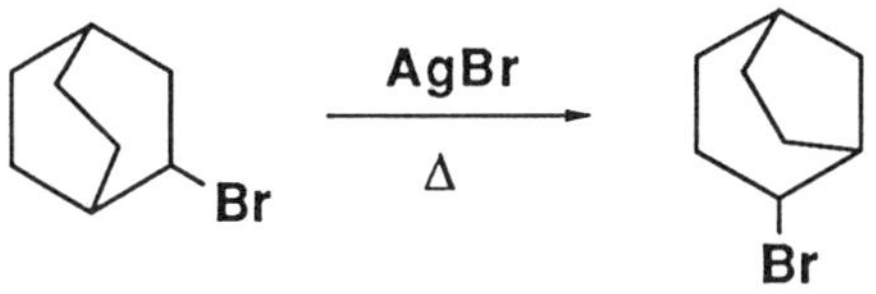

$$\text{(1)}$$

During this period, Doering experimented with a format in which each graduate student was to come up with an original research idea, a forerunner of the proposition system now common in many graduate schools as a requirement for advancement to candidacy for the Ph.D. degree. When my time came I found myself frantically looking through journal articles

*Part of the Doering group about 1947–1948, before I joined the group, on a picnic excursion to Bear Mountain. From left: unidentified head, Milton Farber, Jerry Berson, Mrs. (Bella) Berson, Mortimer Levitz, Al Wolf, Mrs. (Helga) Wolf.*

trying desperately to come up with an original idea. I learned then that creativity cannot be served up on demand. I started then to keep a notebook listing research ideas as they occurred. Sometimes ideas would come while I read papers, sometimes while I listened to someone give a seminar, and sometimes out of the blue with no apparent rationale. But they always came from a background of extensive knowledge and often by relating experience in one area to that in a different area. I found such a notebook of research ideas important in my early career when I was just starting research programs. It became less important later when programs had become established and when a research group was already in place to act on new ideas with less time delay.

One point I should mention about graduate school at Columbia in those days was that the students paid the bills. I bought most of the chemicals I used and paid for glassblowing expenses. We checked equipment such as flasks and beakers out of the stockroom and returned them when finished—clean, very clean. If a flask broke or became uncleanable, we paid. Bags of cracked ice cost us six cents each. Fortunately for me, and other veterans, the G.I. Bill paid for most of these expenses.

## AEC Postdoctoral Fellowship, MIT, 1951–1952

During my final year at Columbia, I was able to think up a research project that served as a suitable basis for an Atomic Energy Commission (AEC) Postdoctoral Fellowship that I was awarded for 1951–1952; the research was to be conducted in the laboratories of Professor John D. Roberts[°] at MIT. While I took the train to Boston to look for housing, a better research idea came suddenly to mind: the idea that one should be able to prepare an optically active primary alcohol of the type RCHDOH, by partial asymmetric reduction of a deuterioaldehyde, RCDO. If such an alcohol, chiral by virtue of hydrogen–deuterium asymmetry, had measurable optical activity, it would be useful for studying the stereochemistry of reactions at primary centers.

The idea had obvious roots. Both Eliel[12] and Alexander[13] had recently published preparations of deuterated hydrocarbons having optical rotations of the order of tenths of a degree. We saw in the example of Al Wolf's research that deuterium was no stranger to Doering's laboratory. Stereochemistry was an important tool in the study of reaction mechanisms in the 1930s and 1940s and was involved in one way or another in much of Doering's research. Optically active secondary alcohols were common at that time, and they and their derivatives were widely used in the study of the stereochemistry of reactions of many types. I was well versed in this literature through my own graduate research on stereochemistry. My idea on the train was a natural extension to the primary carbon of what I had spent several years doing with the tertiary carbon, and Dick Young's results on partially asymmetric Meerwein–Ponndorf–Verley reductions of ketones should provide a natural extension to aldehydes. All of the antecedents for my research idea were in place, and it took only a modest intuitive jump. Such intuitive jumps, however small, are by no means guaranteed or common, and they can provide some of the most exquisite feelings of pleasure and satisfaction given to mankind in general and to scientists in particular.

*Jack Roberts in 1975.*

Professor Roberts thought the new project to be a good one, and he offered strong encouragement. The AEC allowed me to change the original project, and I proceeded to do so but with a further significant change. Bill Saunders° was also a postdoctoral student of Roberts at that time, and he suggested that I consider the use of the magnesium alkoxide instead of the aluminum salt on the basis of some older work.[14] He was right; the use of the magnesium alkoxide was much easier to implement than the Meerwein–Ponndorf–Verley procedure. I resolved some 2-octanol in the standard way to obtain the optically active alcohol, converted it to the magnesium salt by treating with a Grignard reagent, and allowed the mixture to react with butyraldehyde-$d$ (eq 2).

## A Lifetime of Synergy with Theory and Experiment

$$CH_3(CH_2)_2CDO \qquad\qquad CH_3(CH_2)_2CHDOMgBr$$
$$+ \qquad\longrightarrow\qquad + \qquad (2)$$
$$CH_3(CH_2)_5CH(OMgBr)CH_3 \qquad\qquad CH_3(CH_2)_5COCH_3$$

The equilibrium is well on the product side of this equation because ketones are several kilocalories per mole more stable than related aldehydes. The 1-butanol-1-*d* produced was converted to the solid hydrogen phthalate and purified by crystallization. Gas chromatography did not exist in those days, and I could not afford to have any trace of the original 2-octanol contaminating the product because it was expected to have much higher optical activity than my product. After several crystallizations the hydrogen phthalate was hydrolyzed back to the butanol. I remember how nervous I was filling the polarimeter tube and my anxiety in taking that first polarimetric measurement. The rotation was expected to be low, and the measurement was made on the pure liquid in long, narrow-bore polarimeter tubes. Optical activity was measured visually in those days because the human eye is far better at comparing relative light intensities than any photocell

William H. Saunders, Jr., in 1953. We overlapped as postdoctorates with Jack Roberts at MIT. Bill took his Ph.D. with Charles D. Hurd at Northwestern University in 1952, after graduating from The College of William and Mary in 1948. His entire academic career has been at the University of Rochester, where he is now Emeritus. His research has been primarily in elimination reactions and isotope effects.

then available. But the compound did have useful optical activity (optical rotations, $\alpha_D$, measured on the neat liquid in a 4-dm tube were 0.1–0.2°), and my first independent real research project was a success.[15]

Nevertheless, it was still necessary to demonstrate that the optical activity observed was a property of the deuteriobutanol. In the absence of today's analytical and spectroscopic tools one had frequently to use imaginative approaches. I made use of an earlier study of Le Roux and Sugden[16], who obtained the rate constant for bromide exchange between *n*-butyl bromide and the lithium salt of a radioisotope of bromide ion in 90% aqueous acetone (eq 3).

$$Br^{*-} + n\text{-BuBr} = Br^- + n\text{-BuBr}^* \qquad (3)$$

The reaction of my optically active butanol-1-*d* with phosphorus tribromide gave optically active 1-bromobutane-1-*d*. The rate of racemization of this compound with lithium bromide was studied under the same conditions as the radiobromide exchange. The two rate constants agreed within experimental error on considering that each displacement reaction involves complete inversion of configuration. This agreement not only established the optical activity as a true property of the isotopic chiral center but also extended to the primary carbon the historic experiment of Hughes, Juliusberger, Masterman, Topley, and Weiss.[17] This group found that the rate of racemization of optically active 2-iodooctane with iodide ion in acetone is just twice the rate of radioiodide exchange, thereby showing conclusively that each act of displacement resulting in exchange is accompanied by a complete inversion of configuration in which one enantiomer is converted to the other.

I was clearly fortunate to have had the opportunity to work with two such leaders in modern organic chemistry as Doering and Roberts.* Of course, both had important influences on an impressionable young man. As a graduate student with Doering I received an intellectual, analytical view of chemistry and a respect

---

* It is singularly appropriate to me that Doering and Roberts shared the 1990 Welch Prize for their pioneering work in physical organic chemistry. I believe I am the only person to have worked with both of these great chemists.

for the control reaction. These points were reinforced during my year with Roberts. Both men ran group seminars in an informal fashion, and I have continued their style in my own group meetings. Neither required periodic written research reports, and I also followed this style for a number of years, to my regret. I now believe it is important for students to get into the habit of writing research reports on a regular basis. In at least one respect the two men did differ. During that period, Doering was a careful, precise, and consequently slow, writer. Roberts wrote quickly but perhaps somewhat less precisely. I hope I have learned from both. Precision in language is vital for accurate communication, but no work can ever be perfect and at some point one has to say, "It's finished!". I now try to go over my students' research reports with them to point out aspects of grammar as well as communication with the hope that they will learn more than just chemistry as part of their research.

I also regularly attended the evening group meetings of Professor Robert Woodward° at Harvard. These meetings were open to others and were well-attended. Woodward frequently put a problem on the board, and we worked feverishly to try to be the first to put up a correct answer. The problems were often of the "road-map" type, in which we had to determine the structures of a series of intermediates in a reaction sequence. Woodward often then emphasized the reaction mechanisms involved. Road-map problems can still be much fun and highly educational.

We had frequently solved such problems in Doering's group meetings, but there was a new feature in Woodward's. The compounds were characterized frequently not just by chemistry and analyses but also by spectroscopic data. The spectroscopic data at that time were infrared wavelengths. Double-beam infrared spectrometers were just appearing at that time; my Columbia dissertation recorded one of the first in that department, a comparison of an optically active and racemic tertiary alcohol, but Harvard was already making extensive use of their instrument. Most students in Woodward's seminar group had memorized extensive lists of structural effects on carbonyl stretching frequencies. Structural natural products chemistry was changing dramatically and irreversibly. Instead of being deduced by logical inference from chemistry, structures were being determined from empirical spectroscopic correlations. I did not find this newer method to be as much fun, and I realized that I would not make my mark in this type of chemistry. This continuing revolution accelerated a few years later with NMR spectroscopy, and chemistry has not been the same since.

# University of California, Berkeley

## Early Years

Early in 1952, I received an offer of an instructorship at the University of California, Berkeley, which I promptly accepted. In those days it was normal to start academic work as an instructor. I spent my first two years at Berkeley in this rank before being promoted to assistant professor. My wife of two years and I arrived in Berkeley that August, and I have remained at Berkeley ever since. I entered a department steeped in the traditions of physical chemistry, and the spirit of G. N. Lewis° was still pervasive.[18] But changes were clearly in the making. Berkeley has a separate College of Chemistry, which then consisted of a single department. During the war years, while Wendell Latimer° was dean, the decision had been made to bring chemical engineering and organic chemistry to Berkeley. Of course, there were always organic chemists at Berkeley who taught organic courses, but organic chemistry at Berkeley tended to be rather physical. Dale Stewart° was a physical organic chemist who was doing kinetic studies of the hydration of alkenes in sulfuric acid when I arrived in Berkeley. Gerald Branch° had written a 1941 textbook on physical organic chemistry[19] in collaboration with Melvin Calvin°. Calvin's organic chemistry was originally highly physical in nature; he had collaborated with Lewis on the spectroscopy of organic

---

° *See* Appendix A for brief biographical information about each person designated with the ° symbol.

*With my first wife, Mary Ann, walking on campus with our first baby, David, in early 1954.*

compounds and nature of the excited state. After the war, he started work with $^{14}$C, which was becoming more available at that time. Axel Olson° was a physical analytical chemist who did research in physical organic chemistry. He was an early pioneer in displacement reactions and their stereochemistry. My interaction with him was limited because he retired in 1954 and died that same year.

Mainline organic chemistry at Berkeley really began with the arrival of James Cason° and William Dauben° in 1945. They were followed shortly thereafter by Henry Rapoport° in 1946 and by Donald Noyce° in 1948. By the time I arrived, this group had established a beachhead. Courses and policies had been settled, and I benefited from their earlier work. And great things were going on. Calvin had started his now classic work on photosynthesis. Cason, Dauben, and Rapoport emphasized synthetic chemistry, including natural products chemistry. Dauben started his pioneering studies in organic photochemistry. Noyce, Jensen° (who

# A Lifetime of Synergy with Theory and Experiment

*At the old chemistry building in January, 1961. From left: Andrew Streitwieser, James Cason, Ignacio Tinoco, and Donald Noyce. Jim and Don were my colleagues in organic chemistry for many years. Ignacio Tinoco is a biophysical chemist who was Chair of the Chemistry Department from 1979 to 1982. (Picture taken by Prof. William Jolly.)*

arrived in 1955), and I emphasized reaction mechanisms and physical organic chemistry. This balanced group developed organic chemistry at Berkeley to a status that I believe most observers would rank among the best in the world. It was slow at the beginning. Graduate admissions in the early years amounted to barely one new student apiece. Much important early research was done by undergraduates. We worked in old facilities. Most of the organic laboratories were housed in the "Old Chemistry Building", one of the original campus buildings of the 19th century. Rats actually lived in these buildings, and one building annex was even commonly referred to as the "rathouse". Precipitates left to dry had to be covered to prevent contamination from

*Organic faculty luncheon meeting at the Faculty Club, March, 1983. From left: William Dauben, Melvin Calvin, Don Noyce, Fritz Jensen, Andrew Streitwieser, Paul Bartlett.*

debris dropping from the rafters. This situation changed dramatically when we moved to the brand-new Latimer Hall in 1963.

We were a cohesive group that met together regularly. We had an organic faculty luncheon at the Faculty Club every week and still do; this mechanism provided a regular forum for discussing students, courses, teaching, facilities, and research. But we were also part of a complete department. We communicated frequently with our physical chemical colleagues, and I especially received a great deal of knowledge and encouragement from them.

## Stereochemistry of the Primary Carbon

My early experimental work at Berkeley was a continuation of the research on the stereochemistry of the primary carbon that I started at MIT. This work was taken up by a graduate student, William D. Schaeffer, who was the first student to complete a

Berkeley dissertation under my direction. Bill worked at the Union Oil Company in Brea, CA, after getting his Master's degree from the University of Redlands. He was at Brea for only a short time before realizing that his opportunities were limited without a Ph.D. He took a leave of absence and came to Berkeley for further graduate work. He chose to work with me, although I was the younger, probably because this research problem suited his needs. Bill was an excellent and productive student who made a large quantity of optically active 1-butanol-1-$d$ and used it for a variety of studies of reaction mechanisms. His work resulted in seven papers. He got his Ph.D. quickly and returned to Union Oil where, a year later, he was promoted to group leader!

In his solvolysis studies the small amount of racemization found for the reaction of 1-butanol-1-$d$ $p$-nitrobenzenesulfonate with hot acetic acid was increased in the presence of dioxane.[20] The possible intervention of oxonium salts was confirmed by another early graduate student, Sam Andreades, who ran the acetolysis reaction in the presence of dibutyl ether.[21] The optically active 1-butanol-1-$d$ acetate produced was accompanied by a significant amount of undeuterated butyl acetate that arose from reaction of the sulfonate with the ether to give a tributyloxonium ion, $(CH_3CH_2CH_2CH_2)_2O^+CHDCH_2CH_2CH_3$, as an intermediate that reacts further with the acetic acid present to give largely undeuterated ester. These results unexpectedly complemented my graduate research.

In continuing the work of Harold Zeiss at Columbia, I had shown that the solvolysis of optically active 3,5-dimethyl-3-hexyl hydrogen phthalate in methanol gives the corresponding methyl ether with 60% net inversion of configuration (40% racemization). The presence of some solvent addends, such as dioxane and acetonitrile, gave an increased amount of racemization. This result was attributed to the ability of such solvents to solvate the carbonium ion intermediate leading ultimately to an increased fraction of symmetrically solvated intermediates and racemization (Scheme IV).[22,23] An actual oxonium ion intermediate analogous to the tributyloxonium ion of Scheme IV was not established in the dioxane–methanol solvolysis, but a related intermediate is apparent with the use of acetonitrile. One product of the acetonitrile–methanol solvolysis is $N$-(3,5-dimethyl-3-hexyl)acetamide,[22] an expected product of the reaction of a carbocation with acetoni-

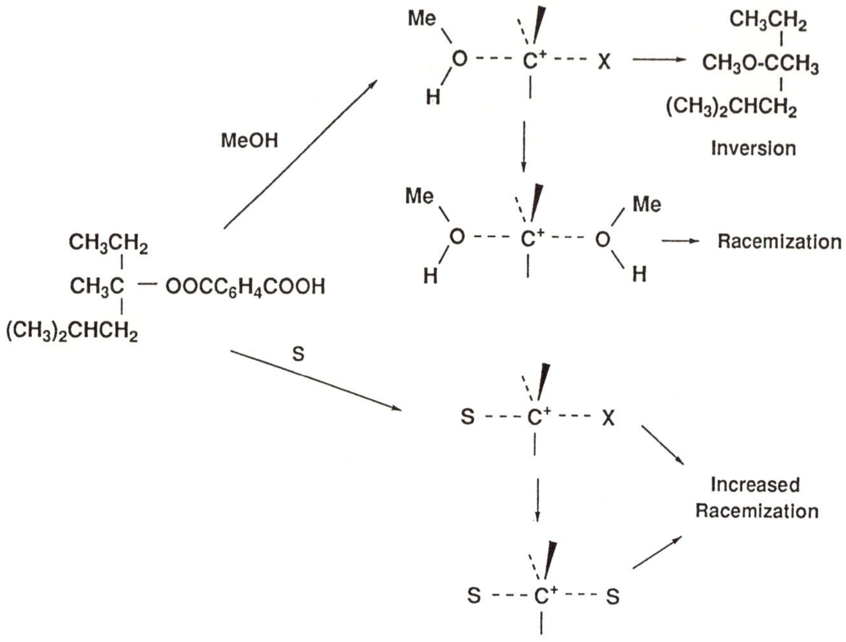

*Scheme IV.*

trile. A clear analogy exists between this result and the Ritter reaction (eq 4), a reaction of tertiary alcohols, acetonitrile, and acid:

$$\text{ROH} \xrightarrow{H^+} [R^+] \xrightarrow{CH_3CN} R-N^+ \equiv CCH_3 \xrightarrow{H_2O} RNHCOCH_3 \quad (4)$$

Thus, the results of Schaeffer and Andreades together with my graduate work showed that both in solvolyses of tertiary systems, in which carbocation intermediates are involved, and of primary systems, which are much more $S_N2$-displacement reaction type in character, nucleophilic solvent addends, such as ethers, can compete with the solvent to produce additional intermediates in the path toward final products.

Bill Schaeffer used his optically active butanol to prepare optically active 1-aminobutane-1-*d* by reaction of the *p*-bromobenzenesulfonate ester with sodium azide followed by reduction with lithium aluminum hydride.[24] One use made of the amine was a study of the amine–nitrous acid reaction. The reaction of primary alkylamines generally gives a variety of products through elimination, solvolysis, and rearrangements, and, accordingly, is not generally a useful reaction for synthesis. In the 1950s, however, it received much attention because of the puzzling variety of products. These products are generally those of "carbonium ion"[*] intermediates, and the reaction sequence was generally written as eq 5:

$$RNH_2 + HONO \rightarrow [RN_2^+] \rightarrow R^+ \rightarrow products \quad (5)$$

In general, the diversity of products and particularly rearrangements are more extensive than products derived from other carbonium ion reactions such as, for example, solvolysis reactions or reactions of halides with Lewis acids. Thus, the carbocations derived from decomposition of alkyldiazonium ions were described as "hot", "high-energy", or "unsolvated" carbonium ions.[25-32] The reaction of $n$-butylamine with nitrous acid in acetic acid gives the expected mixture of products: 1- and 2-butenes, $n$-butyl acetate, $sec$-butyl acetate, and $n$-butyl nitrate. However, our use of optically active 1-aminobutane-1-$d$ gave the result that the 1-butanol-1-$d$ acetate produced was only 31% racemized; it was produced with 69% net inversion of configuration.[33] This high degree of direct displacement reaction product combined with the wealth of typical carbonium ion products led me to a different reaction mechanism, one in which the branch point for alternative reactions is not a carbonium ion, "hot" or otherwise, but rather the alkyldiazonium cation itself.[34] The central idea is that the loss of nitrogen from the diazonium cation is so exothermic and the activation energy so low that the energy range for competing reactions is compressed. That is, the reactions shown in Scheme V can occur at comparable rates because of this compression of the activation energy scale.

---

[*] The term "carbonium ion" is used in its historical context as a generic term for alkyl cations. George Olah has suggested by analogy to ammonium ions that "carbonium ion" be a generic term for alkanonium ions of the type $CH_5^+$ and that alkyl cations be referred to as "carbenium ions". The general acceptance of his suggestion has resulted in "carbonium ion" being now an ambiguous term of uncertain meaning, and I now try to avoid its use by referring to specific "alkyl cations" or "alkanonium ions"; however, in this history the term "carbonium ion" will be used in its historical sense as the cation produced by loss of water from a protonated alcohol (a "carbinol" in nomenclature still common in the 1950s and 1960s).

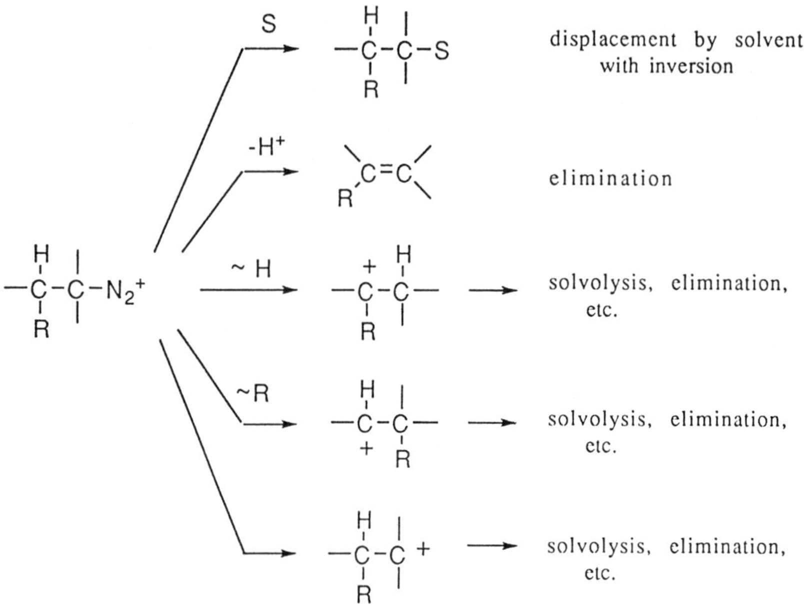

```
                    S        H  H
                   ───→   ─C─C─S          displacement by solvent
                          │  │               with inversion
                          R

                   -H⁺       \    /
                   ───→       C=C          elimination
                             /    \
                            R

  H  H                       H  H
  │  │       ~H              +  │
 ─C─C─N₂⁺   ───→           ─C─C─          solvolysis, elimination,
  │  │                       │  │            etc.
  R                          R

               ~R            H  H
              ───→          ─C─C─          solvolysis, elimination,
                             +  │             etc.
                                R

                             H  H
                            ─C─C +         solvolysis, elimination,
                             │  │             etc.
                             R
```

*Scheme V.*

The argument used was the following. Activation energies for typical solvolysis reactions are on the order of 25–30 kcal/mol. If a competing rearrangement has an activation energy 10% higher, the energy increment of about 3 kcal/mol means that it will represent only about 1% of the total reaction. In the amine–nitrous acid case, if the main reaction has an activation energy of only about 5 kcal/mol, the competing reaction with an activation energy 10% higher now has an activation energy of only 5.5 kcal/mol and would constitute about 30% of the total reaction. This argument illustrates what was meant by a compression of the activation energy scale. In general, if the activation energy of a given process is low, it is likely that the activation energies of competing processes will also be low.

A recent reinvestigation has shown that the racemization we observed actually occurred in the preparation of the amine and not during the diazotization step; the diazotization reaction goes with complete inversion.[35] This result does not change the important conclusions of the proposed mechanism. Except for the addition of ion-pair intermediates and reactions, this mechanism is still, so far as I know, a valid one. Indeed, some rather recent results give the "competing processes" mechanism additional credence as compared with the "hot carbonium ion" mechanism. Recent molecular orbital calculations of alkyldiazonium ions indicate that the loss of nitrogen to give the carbonium ion is not highly exothermic and is more likely slightly endothermic.[36] In this case the intermediate carbonium ion will certainly not be "hot", but the activation energies for forming carbonium ions will be much lower than normal solvolytic processes and the "competing reactions" alternative is still feasible.

One of the significant applications of optically active deuterium compounds to reaction mechanisms shows how the best-laid plans can go awry and give totally unexpected results. During this early period at Berkeley, I sought a suitable example that could show the degree of displacement character in a typical Friedel–Crafts alkylation reaction by study of the stereochemistry of a primary alkyl system with H–D asymmetry. Finding a suitable system was a problem. One could not use an alkyl halide because it would racemize independently through the action of the Lewis acid and the HX produced in the reaction. For this reason we studied the alkylation of benzene with alcohols and boron trifluoride. The reaction can give good yields of alkylation product, but secondary alcohols give extensive rearrangement indicative of carbonium ion intermediates that are essentially free.[37,38] For example, 2-pentanol and 3-pentanol on reaction with benzene and $BF_3$ at 0 °C were found to give identical mixtures of 65% 2-phenylpentane, 25% 3-phenylpentane, and 10% *tert*-pentylbenzene. Thus, rearrangement (eq 6) is fast compared to alkylation.

$$[\text{C-C}^+\text{-C-C-C} \rightleftharpoons \text{C-C-C}^+\text{-C-C} \rightleftharpoons \text{C-C-C-C}^+\text{-C}]$$

⬡ | slower

C-C-C-C-C  +  C-C-C-C-C
    |              |
    Ph             Ph

(6)

Even primary alcohols generally give rearrangement products indicative of secondary and tertiary carbonium ion intermediates. No primary alkyl alkylation product was found. Ethanol, which cannot rearrange to a secondary carbonium ion, was found to be unreactive. Thus, this alkylation process also is not amenable to stereochemical study; however, H–D asymmetry was used to show the stereochemistry of alkylation by a secondary alcohol. Isopropyl alcohol alkylates benzene readily under the action of boron fluoride and cannot undergo hydrogen rearrangements among secondary cations. Peter Stang°, who has since had an outstanding independent research career, showed his early promise as a graduate student in making use of optically active $CD_3CHOHCH_3$ as prepared earlier by Mislow et al.[39] Reaction with $BF_3$ and benzene gave the corresponding deuterated cumene with more than 93% racemization, thus showing that the carbonium ion intermediate in this Friedel–Crafts alkylation is essentially a free and achiral cation.[40]

A possible system that could show the relative roles of displacement on primary centers compared to involvement of primary carbonium ions appeared to be that of the disproportionation of ethylbenzene. In the presence of a strong Lewis acid, such as aluminum bromide or gallium bromide, and HBr, ethylbenzene disproportionates to benzene and diethylbenzenes. Earlier studies led to a mechanism in which one aromatic ring displaces another in protonated ethylbenzene (Scheme VI).[41,42]

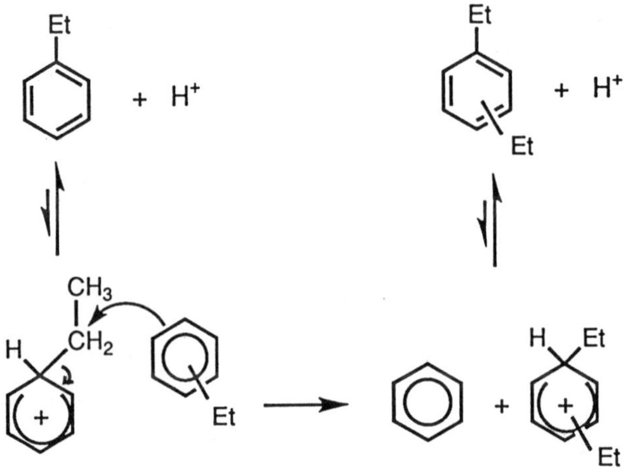

*Scheme VI.*

Our approach was to use optically active ethylbenzene-α-*d* labeled with $^{14}C$ in the ring and to follow its reaction with gallium bromide and HBr in benzene. The loss of $^{14}C$ label would give the rate of aromatic ring exchange, which could be compared with the rate of racemization. If the reaction were of the $S_N2$ displacement type, then racemization would occur twice as fast as exchange exactly as in the Hughes, Juliusberger, Masterman, Topley, and Weiss experiment[17] discussed already. Our experiment showed that loss of $^{14}C$ label was equal to the rate of racemization. However, further study of the deuterium content of the recovered ethylbenzene showed a progressive disproportionation to undeuterated ethylbenzene and ethylbenzene-α,α-$d_2$. These results are consistent with an entirely different mechanism, one that starts with protonation of trace amounts of styrene present to α-phenethyl cation followed by a rapid reversible alkylation of benzene to give 1,1-diphenylethane. The α-phenethyl cation also and more slowly can accept a hydride or deuteride from ethylbenzene-α-*d*. This sequence explains the racemization, loss of $^{14}C$, and deuterium scrambling (Scheme VII).[43–45] Thus, this study that was started to delve into details about one mechanism instead established a totally different one.

This research was done in 1956–1960 by my first female graduate student, Liane Reif (now Reif-Lehrer). She was a vivacious young woman as well as an excellent student, and she always had a ready smile and enthusiastic personality. Liane's brother, Fred, was a Professor of Physics at Berkeley from 1960 to 1990 and is now at Carnegie-Mellon University. Only much later did I learn of the hazards that they faced in their early life. Liane, her brother, and mother, as refugees from Austria, were among the 936 passengers of the *St. Louis,* attempting to flee Nazi Europe

*Ph* ≡ $^{14}C$-labeled          * ≡ optically active

$Ph\overset{*}{C}HDCH_3 \longrightarrow PhC^+DCH_3 \underset{}{\overset{PhH}{\rightleftharpoons}} PhCHDCH_3 + H^+$
                                                                    |
                                                                    Ph

$PhC^+HCH_3 + PhCD_2CH_3$           $Ph\overset{*}{C}HDCH_3$         ↕
    or                              ⟵                    $PhC^+DCH_3 + PhH^+$
$PhC^+DCH_3 + PhCHDCH_3$

*Scheme VII.*

*Liane Reif (now Reif-Lehrer) in her laboratory about 1959. Liane was my first female graduate student.*

in May, 1939. The *St. Louis* was turned away from Cuba and the United States and forced to return to Europe. Most of the passengers did not survive the war. Liane and her mother and brother did succeed ultimately in getting to the United States through the help of her aunt, Lena Klinghoffer. Lena was the mother of Leon Klinghoffer, who was shot by terrorists during the hijacking of the ill-fated *Achille Lauro* (it sank as a result of a fire in 1994). Leon

Klinghoffer is the eponymous subject of a new opera by the Berkeley composer John Adams; my association with Liane gave me added appreciation when my wife and I attended a performance of "The Death of Klinghoffer" during the 1992 San Francisco opera season.

In the 1950s, society had definite ideas of the proper place for women. Liane, however, had the firm resolve that being a woman was not at all inconsistent with being a Ph.D. scientist. I never gave a second thought that there might be anything unusual about such an idea because I had known a number of such women in the Westinghouse Science Talent Search and at Columbia. Perhaps the spirit of the times is indicated by her first job offer from a major chemical company as a Ph.D. chemist—in their library! With her new Ph.D., however, Liane went to Boston, where she lives with her husband, Sam Lehrer, whom she met at Berkeley and who received his Berkeley Ph.D. in physical chemistry. Liane was on the faculty at Harvard Medical School from 1966 to 1985 doing research in the Department of Ophthalmology before leaving to start her own consulting business. If ophthalmological biochemistry seems far from her graduate work, it only shows the value of a broad education in physical organic chemistry.*

During my first year at Berkeley, I learned that Professor Frank Westheimer's° research group at Harvard had prepared chiral ethanol-1-*d* by NADH reduction. The amount obtained was too small to observe polarimetrically, but by a clever series of experiments they showed that the ethanol-*d* obtained was chiral and was, moreover, optically pure.[46] Later, a sufficient amount was obtained to measure polarimetrically.[47] The use of hydrogen–deuterium asymmetry at primary functional centers has since become a standard tool for the study of reaction mechanisms, particularly in biochemical systems, but the desired compounds are now almost always obtained enzymatically. Nevertheless, the optically active material available by the partially asymmetric reductions was useful for establishing the absolute configurations of a number of deuterated centers.[48–50] We made use of this tool

---

* Professor M. J. S. Dewar once claimed to me at a dinner at my house, "Physical organic chemistry is the best preparation for *anything*." There is considerable merit in this statement.

ourselves during early studies on carbon acidity (*see* section titled Kinetic Acidities in Cyclohexylamine) to show that the hydrogen isotope exchange reaction of optically active ethylbenzene-α-*d* with lithium cyclohexylamide (LiCHA) occurs largely with retention of configuration.

## Solvolytic Displacement Reactions

During my second year at Berkeley (1953), I gave a special seminar course on "Modern Aspects of Organic Reaction Mechanisms". During my graduate studies at Columbia and my postdoctoral year at MIT, I had maintained extensive files on the literature of displacement reactions and related solvolysis chemistry. I made ample use of these files during the seminar course, and Bill Dauben suggested that I write a review of the subject for *Chemical Reviews*. He was aware, more than I was in those days, that such a review, if successful, would help my case for tenure. In the 1950s far fewer reviews of various sorts or monographs that summarize various areas of organic chemistry were published than are published today. Indeed, extensive reviews were rare, and the principal forums for reviews were *Chemical Reviews* and *Organic Reactions*.

The timing was propitious. I had had several years to think about the subject, and my own research group was just starting to form. A year later the growing research group would have taken too much time to allow the preparation of such a review. It was also an exciting time in the field. Displacement reactions on alkyl halides are among the most important reactions in organic chemistry. Although solvolysis reactions are synthetically far less important, they are closely related to displacement reactions, and many studies in the 1930s and 1940s were devoted to understanding the $S_N1$–$S_N2$ dichotomy and the role of carbonium ion intermediates. It is difficult to believe now that at that time the very concept of "carbonium ions" was still in dispute, and their nature certainly was. Saul Winstein's° neighboring group effect had become established as a generalized effect in organic chemistry, Don Cram° had recently published his landmark work on phenonium ion rearrangements, new stereochemical results were in

# A Lifetime of Synergy with Theory and Experiment

*My friend and colleague, Bill Dauben, at a surprise birthday party thrown for him by his research group about 1970. The words "Get to work, Jeff!" were subsequently added by one of the group's pranksters, referring to the editor of this series, and the photo was put on Latimer Hall's eighth-floor bulletin board. (Photo courtesy of Jeff Seeman.)*

progress, and the use of strained bicyclic systems initiated the beginnings of the norbornyl controversy. Writing the review took about a year of effort. Toward the end I had that marvelous feeling of knowing that I had created something good. The influence of Doering and Roberts is readily evident in the final product. But the influence of many others is also clearly apparent: notable are Paul D. Bartlett°, whose seminars at Harvard I attended as Jack Roberts's guest; Winstein and Cram, whom I heard and with whom I spoke at meetings; and my colleague Don Noyce, whose research group meetings I shared during my early years at Berkeley.

That issue of *Chemical Reviews* contained only two works: my review, "Solvolytic Displacement Reactions at Saturated Carbon",[23] and an extensive review of allylic chemistry by de Wolfe and Young.[51] The editor, Ralph Shriner, anticipated an unprecedented request for this issue and ordered a large extra printing. The two reviews did produce an extensive demand, and the editor's decision was shown to be fully justified. Several years later, I added some supplementary material bringing some topics up to date, and my review was republished as a monograph by McGraw-Hill in 1962.[52] The two versions of my review turned out to be quite popular and were frequently referred to; indeed, for a number of years they were among the most cited references in chemistry.

Any extensive personal review of the literature of a field is bound to produce new insights and ideas. In the course of the wide reading required, going back even to background material for my dissertation, I had kept extensive notes on $S_N2$ displacement reactions and rate constants. These were distilled into sets of average reactivities for attacking nucleophiles, leaving groups, and central alkyl substrates. The latter set, given here as Table I, has been widely used and even now may be found in textbooks. There have been relatively few reviews even now of the effect of structure on reactivities in $S_N2$ reactions; this gap is rather surprising considering the plethora of reviews and monographs in the current literature and the importance of this reaction in organic synthesis.

Another important insight came from the recognition that the polar substituent constants, $\sigma^*$, proposed by Taft° could be applied to solvolytic displacement reactions. The Taft equation (eq 7) provides a quantitative treatment of inductive effects; $\rho^*$ is a measure of the sensitivity of a given reaction to the effect of each substituent.

Table I. Average Relative Rates of Alkyl Systems in $S_N2$ Reactions

| Alkyl Group | Relative Rate |
|---|---|
| Methyl | 30 |
| Ethyl | 1.0 |
| Propyl | 0.4 |
| Butyl | 0.4 |
| Isopropyl | 0.025 |
| Isobutyl | 0.03 |
| Allyl | 40 |
| Benzyl | 120 |
| Neopentyl | $10^{-5}$ |

SOURCE: Reproduced from reference 23.
Copyright 1956 American Chemical Society.

$$\log(k/k_0) = \sigma^*\rho^* \qquad (7)$$

In eq 7, $k_0$ is the rate constant of the unsubstituted compound, and $k$ is that of the analogous compound containing the substituent. Applied to solvolysis reactions, the magnitude of $\rho^*$ indicates the amount of positive charge developed in the transition state at a developing carbonium ion center. Sufficient data were available in the literature to test this application.[53] For example, solvolysis in ethanol of primary alkyl tosylates having polar substituents on the alkyl chain gave a correlation by the Taft equation with $\rho^* = -0.72$, a rather low magnitude that reflects the $S_N2$ character of this solvolysis with little effective positive charge developed at the central carbon during the reaction. By contrast, for solvolysis in acetic acid of secondary alkyl brosylates, $\rho^* = -3.49$, a much larger magnitude that shows the greater carbonium ion character for the transition state of this reaction. The approach has been useful in characterizing the quantitative values of various substituent effects in carbonium ion reactions.

Molecular orbital concepts were much in my thoughts even while working on the solvolysis review, and a number of orbital interpretations were incorporated in it. Indeed, one of the important aspects of the final review was undoubtedly its use of orbital diagrams. This type of application of molecular orbital (MO) theory in a qualitative sense was just emerging at that time in the literature, and my review was probably the first that made extensive use of such orbital interaction concepts. It was this conceptual approach to chemistry rather than any specific chemical results in the literature that was the most important influence that the work on this review had on my own subsequent chemical research. Carbonium ion reactions themselves became progressively less important in my research after this review. For example, in the review I of course did discuss much of the "nonclassical carbonium ion" literature of the time, but I never did much work myself in this area and never got personally embroiled in the "norbornyl cation" controversy[54], in part because I had no ideas sufficiently novel to warrant my entry into such a crowded field and in part because my involvement with MO theory took me to carbanion chemistry.

Many other physical organic chemists did spend a great deal of effort on the nonclassical carbonium ion question, and they contributed beautiful and creative experiments. The basic question could be stated simply as "are nonclassical carbonium ions of the type involved in rearrangement reactions of strained polycyclic systems actual intermediates or only transition states?". With the fundamental limitation of having to work with real compounds, the question is difficult to answer in any clear-cut and unambiguous manner, and only recently have such nonclassical cations been shown definitively to be capable of existence as intermediates. The tragic aspect is that this question, as fundamental and difficult as it is and despite its stimulating effect on creative new approaches, was not important to the mainstream of synthetic organic chemistry, the part of organic chemistry that drives funding. Accordingly, support for physical organic chemistry diminished and with it the interest of students. Physical organic chemistry is excellent training for many types of industrial research because it involves both synthesis and understanding why compounds react as they do. I believe that the chemical industry in this country has suffered as a result of this declining interest. Only recently, as the intellectual tools of physical organic chemistry have been taken up by organometallic, bioorganic, and some synthetic chemists, is a resurgence of the field becoming apparent.

## Secondary Deuterium Isotope Effects

Deuterium as an isotopic label was much used in the Doering group in the late 1940s and early 1950s. $^{14}C$ was much used as an isotopic label in the Roberts group during the period that included my postdoctoral year at MIT. We have already seen the importance of isotopes as tracers in my study of reaction mechanisms, but there was also the appreciation of kinetic effects in replacing one isotope of an element by another. The differing masses of isotopes affect vibration frequencies and therefore the thermodynamics of compounds and kinetics of their reactions; the effects are greatest for hydrogen because the isotopes deuterium and tritium differ so much in mass. *Primary isotope effects* arise in reactions in which a bond to the isotope is made or broken; *secondary isotope effects* apply when bonds to the isotope are not made or broken but are perhaps only altered. Since their discovery in the mid-1930s isotope effects have been important tools in the delineation of reaction mechanisms. My awareness of this chemistry developed in my own reading of current and past literature. One of the earliest independent research ideas that I recorded in 1952 in my idea notebook was a suggestion that deuterium isotope effects might be used as a gauge of hyperconjugation in carbonium ion reactions.

The central concept was that delocalization of the positive charge of a carbonium ion center to an adjacent C–H bond would result in weakening of the bond. The concept was usually symbolized by "no-bond resonance structures" such as: H–C–C$^+$ ↔ H$^+$ C=C. Such bond weakening should result in an observable secondary isotope effect with the corresponding C–D bond. The magnitude of such an effect would then be expected to provide a quantitative measure of hyperconjugation and could also have conformational and stereochemical consequences. Not long thereafter I learned that Jack Shiner° had independently arrived at the same idea and was already working on it.[55] My disappointment in not being first was tempered by the knowledge that I had thought of a research idea sufficiently good that someone else thought it worthwhile and was pursuing it. That was an important lesson for I realized that if I had one good idea I would surely have another. The lesson has been amply confirmed by my subsequent experience. My late colleague Joel Hildebrand° was fond of pointing out that few people have only one good idea; they have either none or many.

Bob Fahey° was an undergraduate student at Berkeley in the early 1950s and was seeking an undergraduate research problem. Because the β-deuterium isotope effect as a measure of hyperconjugation had already been determined, I suggested that he measure an α-deuterium isotope effect as in the solvolysis of cyclopentyl-1-*d* tosylate. In this reaction a tetrahedral C–H bond changes to a trigonal bond, which should have a stronger stretching force constant. Thus, I expected a small inverse isotope effect. Such inverse isotope effects were rather rare, and I thought it would be useful to document at least one additional example, particularly because such a research project was well suited for an undergraduate project. Bob Fahey's result[56] of a normal deuterium isotope effect of magnitude of $k_H/k_D = 1.15$ came as a complete surprise and necessitated a more detailed analysis of the system. I then realized that one must consider not just the change in stretching force constant during the reaction but also the bending force constants. One of these, corresponding to the out-of-plane bending of the C–H(D) bond, is dominant. The reduced coordination around carbon at the transition state results in a softer bending vibration and a normal $k_H/k_D$ isotope effect. One lesson in this research result is to be wary of the superficial or popular explanation and to seek potential underlying principles that might turn out to be more important. In the case of this isotope effect I had to go back to fundamentals on the effect of isotopes on partition functions. With some simplifying approximations I derived eq 8, which gives the kinetic isotope effect expected for changes in bond frequencies (v) between reaction state and transition state (‡). The factor 0.187 includes the computed mass effect of changing from hydrogen to deuterium.

$$\frac{k_H}{k_D} = \exp\left[\frac{0.187}{T}\sum(v_{H_i} - v_{H_i}^{\ddagger})\right] \qquad (8)$$

It became immediately clear that the magnitude of this isotope effect can provide an important measure of the structure of a transition state. In effect, by this approach the isotope effect gives a measure of the infrared spectrum of a transition state. The looser

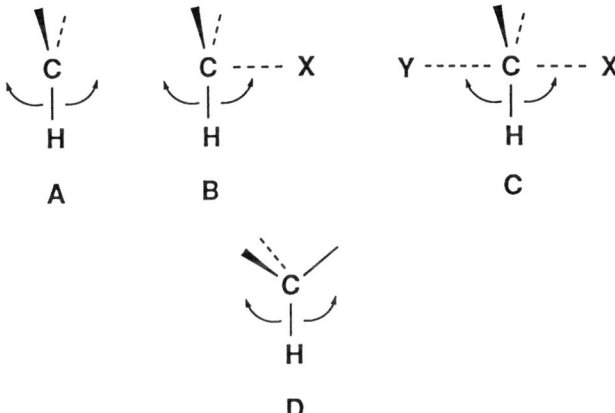

*Figure 2. "Out-of-plane" C–H bending motions in various transition states. As coordination around carbon increases from structure A to structure C compared to the tetrahedral precursor, D, the bending motion is impeded and the bending frequency increases, which leads to a weaker or even inverse isotope effect.*

the C–H(D) vibration in the transition state, the greater the isotope effect (Figure 2).[57]

In an $S_N2$ reaction the entering and leaving groups are relatively close to the reacting carbon, and the bending motion of the H(D) is impeded; the result is only a small net change in the bending frequency and a small isotope effect, with $k_H/k_D$ close to unity. Even in the solvolysis reaction the leaving sulfonate group is still rather close to the reacting carbon, and the bending frequency is not as low as it would be in an isolated carbonium ion. Thus, the isotope effect in the solvolysis reaction is of lower magnitude than expected for an unencumbered carbonium ion.

This interpretation was exciting in its ability to explain existing facts and to provide predictions for other reactions, but I had had to delve into unfamiliar statistical thermodynamics to provide the analysis. I recall consulting with Kenneth Pitzer° to check my analysis and being encouraged by him. I presented the theory at a physical chemistry seminar at Berkeley and had it survive intact from the scrutiny of my expert colleagues.

This work, done early in my Berkeley career, remains one of my most significant contributions. Yet, I clearly had no concept of this significance when the research started. Indeed, the experiments involved in the original research amount only to a few simple preparations and a modest number of kinetic runs. It is sobering to reflect that if I had written a research proposal suggesting these experiments and trying to indicate why they are important I would have gotten it all wrong and undoubtedly would not have been funded! In the more than three decades since this α-deuterium isotope effect was discovered[56], it has been a significant research tool in the study of many reactions in organic chemistry and biochemistry. It has been used, for example, in the study of a number of enzyme reactions with aldehyde derivatives to evaluate the nature of the transition states.[58] In some cases, and apparently in proton-transfer reactions of enzymes, coupling of the transferred hydrogen to other hydrogen motions and tunneling can introduce complications[59], but the approach remains a useful one that is still applied extensively.[60] Ab initio calculations of transition states are now common; such calculations of some $S_N2$ transition states have shown that the stretching force constants cannot be neglected and can lead to inverse isotope effects.[61,62] Nevertheless, a recent thorough study has confirmed the essential correctness of our original treatment[63]. Indeed, the original paper appears to have achieved the status of such a classic that it is frequently no longer referred to directly compared to references to reviews. This example points up a limitation in using tools such as the *Science Citation Index* to quantitate scientific significance!

By the end of my first decade in Berkeley a variety of secondary isotope effects were known as summarized in an excellent comprehensive review by Halevi in 1963.[64] His review covers many examples of anharmonicity effects. Even at that time it had been known that one of the consequences of the anharmonicity of bond vibrations is that, averaged over the time of a vibration, bonds to deuterium are slightly shorter than bonds to protium. This effect causes small changes in dipole moments of deuterium compounds compared to protium isomers and can also give rise to chemical effects. In the late 1950s Halevi showed that deuterated acids, such as $CD_3COOH$, are slightly less acidic than the protium analogs.[65-67] These results were attributed to such anharmonicity effects; the slightly shorter C–D bond puts more electron density closer to the carboxylic acid group and lowers its acidity. The effect was confirmed shortly thereafter, and Bell and Crooks[68] showed that the p$K$ difference between DCOOH and HCOOH agrees with vibrational energy changes; that is, although the fundamental source of the effect is in the anharmonicity of bond vibrations, the changes are implemented by changes in zero point energy in the same way that other isotope effects are manifest. An important difference is that small changes in many vibrations are involved, and an analysis such as that given already for the α-deuterium isotope effects cannot be applied.

We wondered whether these secondary deuterium isotope effects on acidity would show empirical patterns similar to the inductive effects given by normal substituents. The magnitude of the effect is consistent with deuterium isotope effects on dipole moments.[69,70] Harvey Klein took this problem and made careful p$K$ measurements on DCOOH, CD$_3$COOH, (CD$_3$)$_3$CCOOH, and several deuterated benzoic acids. The comparison of acetic and pivalic acids was particularly significant. In CD$_3$COOH the deuteriums are one atom closer than in (CD$_3$)$_3$CCOOH. In the pivalic acid system each deuterium should be 2–3 times less effective if a normal inductive falloff is operative, but because there are 3 times as many deuteriums, the total effect should be substantially unchanged. In fact, the p$K$ difference for the deuterated pivalic acid, 0.018 ± 0.001, compared with that for acetic acid, 0.014 ± 0.001, agrees with this treatment of these isotope effects as normal inductive effects.[69,70]

Further studies, however, showed that treating such isotope effects as a type of inductive substituent effect is oversimplified. Ring deuterium increases the rate of solvolysis of benzhydryl chloride, but the orientation effect (ortho 1.9%, meta 1.5%, para 1.0%) does not accord with typical ring substituent effects.[71] Comparable effects have been reported for the ionization equilibria of deuterated triphenylmethyl chlorides in liquid sulfur dioxide.[72,73] The effects of ring deuteration on the acidity of anilinium ions were shown to be rather small.[74] Similarly, small rate-retarding effects were found for ring deuteration in the LiCHA-catalyzed α-deuterium exchange of toluene, but the orientation effects differ from the carbonium ion results; hence, we concluded that such isotope effects "have limited usefulness as probes of electronic structure".[75]

These effects, although small, are significant. The direction is such as to reduce the expected magnitude of the normal α-deuterium isotope effect discussed already for carbocation-type reactions and equilibria; nevertheless, this "inductive effect" of deuterium is rarely given explicit consideration in applications of isotope effects.

## Molecular Orbital Theory for Organic Chemists

*Molecular Orbital Theory for Organic Chemists* is my first book[76], and it was the outgrowth of an interest in Hückel molecular orbital (HMO) theory that started in graduate school. At one of his group seminars Professor Doering told us of the Hückel $4n + 2$ rule but gave us only a qualitative background. During my postdoctoral year at MIT, Professor Roberts gave an informal series of lectures on HMO theory, which he had just recently learned himself. We learned by doing a number of calculations of compounds of interest in Roberts's research group, and we completed some joint projects that made use of these HMO calculations.[77,78] Doing these calculations and interpreting the results filled me with a sense of power at being able to calculate theoretically an energy quantity for some types of compounds to help understand organic chemistry. Unfortunately, we also understood the many drastic approximations that were made in order to permit such calculating at all even though, at that time, I was not aware of the growing literature in Hückel theory. Many of the recent developments in the 1940s were the work of theoreticians such as Coulson and Longuet-Higgins.[79–81] Many of the applications had been with qualitative aspects of chemistry, but simple Hückel theory had also been shown to give quantitative correlations with such physical properties as bond distances, and ionization and reduction potentials. How well these calculated π-energy differences would correlate with experimental rates and equilibrium constants remained in question.

One of the first opportunities for a quantitative test came in a seminar given by Norman Lichtin° at one of Professor Paul Bartlett's group meetings, on the ionization equilibria of triarylmethyl chlorides in liquid sulfur dioxide (eq 9).[73,82,83]

$$Ar_3CCl = Ar_3C^+ + Cl^- \qquad (9)$$

The equilibria involved triphenylmethyl cations with meta- and para-phenyl substituents and naphthyldiphenylmethyl cations, systems that were ideal for examining π-energy changes only. The fundamental question was whether the sigma bond energy changes are effectively constant for such a closely related series and whether the Hückel method would give meaningful quantitative results. The test involved calculating the π-energies of both sides of the equilibria, $\Delta E_\pi = E_\pi(Ar_3C^+) - E_\pi(Ar_3CCl)$, and comparing the results with the experimental equilibrium constants. The π-energy of the chloride was approximated as the sum of the π-energies of the constituent aromatic rings, and these values were straightforward to compute. The cations, however, are large systems for computations with only an electric calculator. The year was 1951, and electronic computers were not yet generally available. I recall that one calculation required solving a 25 × 25 determinant. Two attempts gave me different answers, and it required a third attempt and a total of about one ream of paper to get the answer. Today, of course, the problem could be solved rapidly even with a personal computer. Nevertheless, the results showed an excellent quantitative correlation between theory and experiment (Figure 3). Even the discrepancy shown by the point off the curve could be explained. This point belongs to α-naphthyldiphenylmethyl cation; steric hindrance at the α-naphthyl position prevents the same degree of conjugation to the central carbon as in the other cations, and this cation is therefore not expected to be as stable as calculated. This work showed that Hückel theory could be applied quantitatively to some kinds of reactions, but it also pointed up some of the potential limitations. It gave me my second independent publication[84] and established some of my early research projects at Berkeley. These projects involved rate and equilibrium measurements of polycyclic aromatic compounds that would be suitable for comparison with simple HMO calculations.

One example was the solvolysis in acetic acid of polycyclic arylmethyl tosylates.[85,86] The departure of the tosylate ion gives the benzylic carbon a substantial degree of carbonium ion character at the transition state (eq 10).

$$ArCH_2OTs \ ---[ArCH_2^+ \cdots OTs^-]^\ddagger ---> \text{Products} \qquad (10)$$

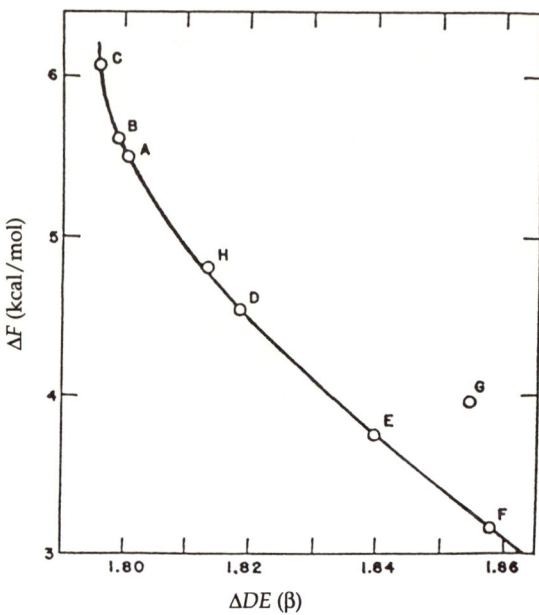

*Figure 3. Correlation curve of experimental ionization free energies (ΔF) for triarylmethyl chlorides in $SO_2$ and Hückel delocalization energy differences (ΔDE, in units of the Hückel parameter β) for the triarylmethyl cations. (Reproduced from reference 84. Copyright 1952 American Chemical Society.)*

The question was whether the same type of HMO treatment that had been applied to the triarylmethyl cation *equilibria* would apply as successfully to carbonium ion *reactions*. The results summarized in Figure 4 show a grouping of the points into two principal linear correlations, one for benzylic- or β-naphthyl-type groups and another for α-naphthyl-type groups. The $\sigma^+$ numbers shown in Figure 4 are composite reactivities for arylmethyl groups derived from our acetolysis studies as well as related reactions run by other research groups. The correlations showed that Hückel theory could be used quite successfully for compounds that are closely related, but limitations of the simple theory are also apparent. The dispersion into two types of correlation was shown subsequently to result from an important limitation of Hückel-type approaches, namely, the failure to account adequately for electron repulsion effects.

A related research topic was that of electrophilic substitution of polycyclic aromatic hydrocarbons. We measured the rates of

# A Lifetime of Synergy with Theory and Experiment

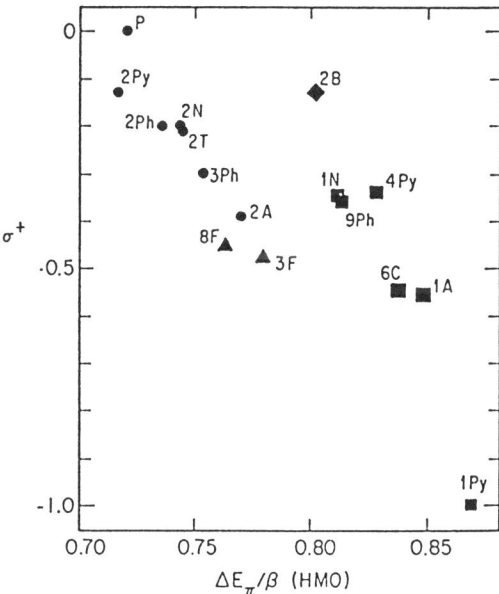

Figure 4. *Correlation of carbonium ion reactivities expressed as composite $\sigma^+$ values with Hückel $\pi$-energy differences ($\Delta E_\pi$) between ArCH$_2^+$ and the $\pi$-system of the reactant, ArH. Symbols: P is phenyl; N is naphthyl; A is anthracyl; Ph is phenanthryl; Py is pyrenyl; T is triphenylenyl; C is chyrsenyl; F is fluoranthyl; B is biphenylenyl; HMO is Hückel molecular orbital. Types of position:* ●, *phenyl or $\beta$-naphthyl;* ■, *$\alpha$-naphthyl;* ▲, *fluoranthyl;* ♦, *biphenylenyl. (Reproduced from reference 86. Copyright 1970 American Chemical Society.)*

replacement of tritium by hydrogen (protodetritiation) in trifluoroacetic acid of a number of hydrocarbons and compared the reactivities with Hückel theory.[87] One of the graduate students working on this project was Adolphus Lewis, my first black coworker, now on the faculty of Howard University.

A few other groups were pursuing similar studies at the same time and for the same purpose. (An extensive review of this work is given in section 12.3 of my MO book.[76]) Notable among these was that of Michael Dewar, whom I have mentioned a number of times previously. These studies revealed a great deal about the role of conjugation at the transition states of various reactions. Professor Dewar has expressed the opinion, with which I heartily agree, that many studies of substituent effects even today would benefit by including conjugating nonpolar substitu-

*Professor Michael J. S. Dewar is one of the leading physical organic and theoretical organic chemists of this century.*

ents, such as in biphenylyl or naphthyl systems. The interactions of a dipolar substituent with a charged reaction center as normally measured by common Hammett σρ studies frequently differ substantially from the purely conjugation effects that could be revealed by the study of different π-systems.

During the course of these early studies I followed much of the literature dealing with MO methods of π-electronic systems and their applications to chemistry. No textbooks were available that were suitable for organic chemists, and a number of suggestions were made that I should write such a book. The opportunity came during my first sabbatical at Berkeley in the 1959–1960 academic year. For family reasons it was necessary for me to remain in Berkeley rather than to spend the usual traveling sabbatical elsewhere. I obtained a National Science Foundation

(NSF) Science Faculty Fellowship that permitted me to take a sabbatical for a full year.

The book started with notes I had taken from Professor Roberts's lectures during my postdoctoral year at MIT. These notes developed extensively over the years and became further organized for a 10-week special course on MO theory that I gave in 1958 at the Shell Development Company in Emeryville, California, at the suggestion of Dave Stevenson°. These notes formed only the bare bones of the book, and substantial further efforts of research, organization, and writing were required to complete the final work. The work of scholarship was helped greatly by my acquisition of one of the first copiers to appear on the market at that time. Unlike modern dry copiers, this one involved a wet development process and the dried sheets tended to curl and had a sulfurous odor, but it was still invaluable to have actual references available at the writing desk. Little did I realize then how we would all become inundated by piles of such copied papers!

My family, which by then included two children, had just moved to a new house that had an unfinished area downstairs. During that sabbatical year I also partitioned off this space into a bedroom, playroom, and bathroom for the children. During morning hours I would do construction work while thinking about the organization of the writing even to the extent of imagining whole sentences and paragraphs. During the afternoon, the material that had been mentally refined in the morning was put to paper together with the necessary research documentation. This combination of some physical activity juxtaposed with writing is well recommended as a way of dealing with a project requiring a substantial intellectual effort. On other occasions during long walks I have been able to mentally organize writing tasks. I suspect that what happens is the result of an alternation of moments of intense concentration with other moments when attention needs to be paid to what one is doing physically (e.g., walking or hammering). During the latter periods the subconscious mind is undoubtedly still active with the material developed during the periods of concentration so that the next period of concentration has the results of the subconscious activity to work with. Perhaps, however, it is simply a matter of getting the blood circulating in the brain. However it works, this

*One of my favorite pictures of my children, David, 6, and Susan, 4 (1960).*

approach has often provided solutions to difficult problems. Of course, this psychology is well known and has been discussed by others, such as the French mathematician Jules-Henri Poincaré.[88]

The book *Molecular Orbital Theory for Organic Chemists*[76] was well received. I believe that the gratifying success of the book was a result of its chemistry orientation. The mathematics of doing simple Hückel theory is there but with numerous applications to chemical equilibria and reactions. This book and an equally successful smaller volume by Jack Roberts[89] were also timely. The importance of the Hückel $4n + 2$ rule was beginning to receive its much belated recognition, and organic chemists were eager to learn more. These books helped to educate a new generation of organic chemists into the use of MO concepts. Critics have generally been kind to my book for its presentation of MO theory as it existed by the end of the 1950s, but it has also been criticized

# A Lifetime of Synergy with Theory and Experiment

*Bill Johnson and Edith and Jack Roberts at a dinner in honor of Professor Johnson at Fisher Island, Miami, during the ACS meeting, April 1989. I have known all three for many years. Bill Johnson sadly died in 1995.*

for not predicting adequately the role of qualitative perturbation approaches that were to become the really important application of HMO theory in organic chemistry. This application makes use of the "frontier orbitals" of Fukui[90,91], the highest occupied and lowest unoccupied molecular orbitals (HOMO and LUMO, respectively) as developed extensively by Dewar.[92] This criticism is largely valid even though the book does contain one of the first applications of an orbital symmetry perturbation approach to an organic problem.

This latter example arose from some solvolysis reactions of methoxy-substituted benzonorbornyl sulfonates carried out by George Wiley,[o] who was then a temporary assistant professor at Berkeley. George's results were not what he expected, and he discussed them with me and wondered if MO theory could provide a rationalization. The simple perturbation treatment not only provided a completely satisfying explanation but required a conjugating effect of the phenonium ion type that was exactly what George had been seeking!

HMO theory is "simple" because of some drastic approximations. It applies only to the π-electrons of a hypothetical framework of carbons all in the same plane and with equal bond distances. The π-electron integral at each carbon is given a common but unspecified value of α. Similarly, the bond integral for all pairs of bonded carbons is given the value β. The method works best for benzenoid hydrocarbons, but we have seen that the theory has been used for other types of compounds as well. If the π-system contains an atom other than carbon, modifications are required in α and β. The simplest way of incorporating these changes is to define new parameters for the heteroatom X in units of β as in eqs 11 and 12.

$$\alpha_x = \alpha_x + h_x \beta \tag{11}$$

$$\beta_{cx} = k_{cx} \beta \tag{12}$$

The $\alpha_x$ is the more important parameter and embodies the difference in electronegativity between X and C; if X is more electronegative than C, then $\alpha_x$ is a more negative energy than $\alpha$ and $h$ has a positive value (because β is a negative energy). In reviewing the literature and doing a number of test calculations, I derived a "best set" of recommended values for $h$ and $k$ for some important heteroatoms for use in the book. A partial list is given in Table II. These values were not recommended for quantitative correlations but simply to show trends. They have, however, been

Table II. Suggested Parameter Values for
Heteroatoms in Hückel Molecular Orbital Theory

| Element X | $h_x$ | $k_{CX}$ |
|---|---|---|
| B | −1 (−0.45) | 0.7 (0.73) |
| N(N·) | 0.5 (0.51) | 1 (1.02) |
| N(N:) | 1.5 (1.37) | 0.8 (0.89) |
| O(O·) | 1 (0.97) | 1 (1.07) |
| O(O:) | 2 (2.09) | 0.8 (0.66) |
| F | 3 (2.71) | 0.7 (0.52) |
| Cl | 2 (1.48) | 0.4 (0.62) |

NOTE: X· contributes one electron to the π-system; X: contributes two electrons.
SOURCE: Values in parentheses are from reference 93.

widely adopted, and a recent analysis has yielded[93] values gratifyingly similar to those proposed in my MO book more than three decades ago.

The experimental rates and equilibrium constants measured during the 1950s and 1960s as quantitative tests of Hückel theory showed the important limitations in dealing with quantitative chemistry of any approximate quantum mechanical procedure that neglects electron repulsion and considers only π-electrons. The real value of Hückel theory is that it gives the correct nodal properties of MOs, and these alone can lead to important predictions and understanding of chemistry. Neither I, nor many of my colleagues who had had equal opportunity, discovered the general concepts now known as the Woodward–Hoffmann rules for pericyclic reactions. But it is probably fair to claim that my book and Jack Roberts's book did help to prepare organic chemists to appreciate these rules when they were discovered only a few years later. I still recommend teaching Hückel theory at the graduate and advanced undergraduate level because it is important for the qualitative understanding of a great deal of chemistry and particularly for the Woodward–Hoffmann rules.

For many years one of my regular teaching duties was a course on physical organic chemistry for senior undergraduates and first-year graduate students. For most of this time the course was known as Chemistry 127 and included MO theory. As soon as the Woodward–Hoffmann rules appeared they were incorporated into the course. I had met Roald Hoffmann° frequently, and we had discussed the value and limitations of the extended Hückel theory that he had used so successfully. When the Nobel prizes in Chemistry were announced in early October, 1981, I received a phone call from the *New York Times* asking for some information for their readers. "What are orbitals?" they asked. They wanted a picture of an orbital for a simple compound, and I immediately thought of the splendid orbital diagrams in Jorgensen and Salem's book.[94] One of the diagrams of a molecular orbital of water would do the job, and I pointed out that they could get a copy of the book from the New York Public Library. But there wasn't time! They had a press deadline. Faxes were not as common then as now, but the University of California Information Office had such a machine. The *New York Times* arranged for a courier to take my copy of Jorgensen and Salem to the University of California Information Office, and a picture was transmitted to New York. Thus, the molecular orbital of water reprinted in the *New York Times* that day actually derived from my personal copy of the book!

This adventure was quickly followed by a phone call from the then editor of *Science*, Phillip Abelson, asking me to write an article about the contributions of Hoffmann and Fukui. October was one of my busiest times coming at the start of the academic year, but this responsibility was not one to be refused. I realized that I was probably one of the most qualified people available for the assignment and completed it despite the stringent deadline involved. The resulting article[95] also cites Dewar's contribution to this research. I have frequently expressed my opinion that Michael Dewar could appropriately have shared this prize.

Two years later, in 1983, I had occasion to do a video interview of Roald Hoffmann for the "Eminent Chemists" series of the American Chemical Society. I was giving a series of lectures at Ithaca College, and we carried out the interview in a studio at nearby Cornell University. In choosing me as the interviewer I doubt that the ACS organizers realized that Roald and I are both alumni of Stuyvesant High School and of Columbia College, just 10 years apart. Roald was also a Science Talent Search trip winner in 1955, again just one decade after my own trip. I had done some homework on Roald's background, of course, for the assignment,

# A Lifetime of Synergy with Theory and Experiment

*Kenichi Fukui (left) visited Berkeley in October, 1972. During his visit, we visited wineries in the Napa Valley. This picture of us at the Sterling Winery was taken by Professor Takeuchi Fueno of Osaka University, who was spending his Sabbatical leave at Berkeley. Professor Fukui shared the 1981 Nobel Prize in Chemistry with Roald Hoffmann.*

but the interview itself was unrehearsed and spontaneous. I think the final product is quite effective in telling Roald's fascinating story.

I was also privileged to meet Kenichi Fukui° on several occasions. We met initially on my first trip to Japan in October, 1971, at the time that I gave a lecture in Kyoto. He and his charming wife entertained my wife and me that evening although I found that I was to do some of the entertaining! We went to one of their favorite local clubs where I as a guest was expected to get up and sing something to the almost wholly Japanese audience. I sang the only song I could think of at the time to which I knew all the words, "Mairsey Doats". Two years later, in October 1973, Professor Fukui visited California, and I was able to take him to

*I was a member of the Executive Board of the Organic Division of the American Chemical Society for 1972–1975, and Chairman for 1973–1974. One of my pleasant duties was to present the Roger Adams Award to Georg Wittig (1897–1987) during the National Organic Symposium, Tallahassee, FL, in June, 1973. Professor Wittig (left) shared the Nobel Prize in 1979 with Professor Herbert C. Brown. This picture was taken in the university studio with my camera, and I developed and printed the picture myself; the official studio portrait was lost!*

*Shuji Ozawa, Sue, and me at Hakone Park with an awesome view of Mt. Fuji in October, 1971. Shuji was a former student who introduced us to Japan. He arranged our stay at a recreation center in Hakone Park for employees of Teijin Ltd. We were the first Westerners to stay at this center where we were introduced to the Japanese hot tub and sleeping on futon. On a later trip to Japan I climbed several stations on Mt. Fuji.*

the Napa Valley for the fall grape harvest, together with Professor Takayuki Fueno, a physical organic chemist who was on sabbatical leave in my laboratory at that time from Osaka University.

That first trip to Japan was a marvelous adventure. We were met and introduced to Japan by my former student, Shuji Ozawa. Shuji showed us the proper use of *ofuro*, the Japanese hot tub. We were so taken by it that six years later we built a cedar hot tub in our patio and have used it almost daily since. We also enjoyed staying at *ryokan*, the Japanese Inn. We took readily to Japanese food, which at that time was not as commonly available in this country as it is now; we still frequently enjoy sushi and sashimi. On that trip we also met Professor Yasuhide Yukawa, who guided us around Osaka and Nara. Professor Yukawa wrote a preface for the Japanese edition of my MO book. His co-worker, Yuho Tsuno, did the translation. I met Professor Tsuno, now at Kyushu University, when I was invited as a plenary speaker at two of the Kyushu International Symposia of Physical Organic Chemistry, KISPOC I and KISPOC V.

*Sue and I were greeted by Professor Yukawa during my first trip to Japan in 1971. Professor Yukawa is one of the most prominent Japanese physical organic chemists and wrote a preface for the Japanese edition of my MO book.*

# A Lifetime of Synergy with Theory and Experiment

*Professors Tsuno (left) and Yukawa (right) during KISPOC I in 1982. Professor Yuho Tsuno translated my molecular orbital book into Japanese. They are perhaps best known for the Yukawa–Tsuno equation, an extension of the Hammett equation for substituent effects.*

# Carbon Acidity

## Kinetic Acidities in Cyclohexylamine

During my early period at Berkeley, conspicuously absent among the reaction systems amenable to simple molecular orbital treatment were systematic studies of carbanions. Only a few studies of carbon acidities of conjugated hydrocarbons were available, and even these were mostly qualitative. My initial work in carbanion chemistry began in the late 1950s and had as its objective quantitative measures of relative acidities of hydrocarbons suitable for testing with possible Hückel correlations. Thus, these studies were directed toward aromatic systems and benzylic-type carbanions having π-electronic systems to which the HMO method could be applied. That is, the idea was to apply to carbanion reactions the types of Hückel correlations that were being applied to carbonium ion reactions. In particular, I thought it important to demonstrate whether the limitations of Hückel theory that were showing up with carbocations would apply as well to the corresponding carbanion systems.

We had first to find a suitable reaction system and to work out the experimental techniques to get such quantitative data. An extensive study of kinetic and equilibrium acidities followed, first with lithium cyclohexylamide (LiCHA) and then cesium cyclohexylamide (CsCHA) in cyclohexylamine as solvent. Earlier work had been done, notably by Jack Roberts[96] and by Shatenstein[97] in Russia, on rates of proton exchange (kinetic acidities) catalyzed by potassium amide in liquid ammonia. Liquid ammonia is not a

*Posing in one of my laboratories in May, 1963, shortly after moving into our brand-new Latimer Hall. The apparatus was that used for our kinetic acidity studies in cyclohexylamine; I wouldn't dream of interfering with one of my student's real experiments. (Reproduced with permission from Time-Life.)*

particularly good solvent for many organic compounds, and its low boiling point makes it inconvenient to use. It was for such reasons that I chose an organic amine as a solvent. Moreover, I wanted to use a primary amine in order to have as large a proton pool as possible in the solvent for isotope exchange studies. Of the readily available primary amines that we studied in the initial work, we found that only cyclohexylamine dissolved its own lithium salt. Cyclohexylamine is also inexpensive and readily purified. This innocent beginning led to extensive studies that are still in progress. Numerous experimental problems had to be solved dealing with the handling of air-sensitive solutions and reactions. LiCHA was prepared by the reaction of cyclohexylamine with ethyllithium, which in turn was prepared from ethyl bromide and lithium metal. Neither ethyllithium nor butyllithium, the common reagent today, was available commercially at that time, and we had to make our own organolithium intermediates. Ethyllithium is a solid that crystallizes readily from benzene; this convenient property has been forgotten by today's organic chemists who automatically reach for the commercially available butyllithium in solution.

My first student in this research area was Dale Van Sickle who, in 1957, developed vacuum-line methods for studying the protodedeuteration of toluene-$\alpha$-$d$ with LiCHA (eq 13).

$$C_6H_5CH_2D + C_6H_{11}NH_2 \xrightarrow{C_6H_{11}NHLi} C_6H_5CH_3 + C_6H_{11}NHD \qquad (13)$$

Studies of the reaction mechanism with kinetics, isotope effects, substituent effects, and stereochemistry showed that the reaction is between toluene and monomeric LiCHA and involves a transition state that has a significant degree of benzyl anion character.[98-102] For example, the reaction has a large primary isotope effect, $k_H/k_D = 12$, which indicates that the C–H(D) bond is extensively broken at the transition state. The secondary isotope effect, measured by the ratio $3k(C_6H_5CH_2D)/k(C_6H_5CD_3) = 1.31$, shows also a substantial decrease in coordination at the benzylic carbon (see previous section, Secondary Deuterium Isotope Effects). $m$-Xylene and $p$-xylene are 0.60 and 0.29, respectively, as reactive as toluene, which indicates the buildup of negative change at the benzylic position. This conclusion was confirmed by a more extensive study of toluene substituent effects. A series of meta- and para-substituents follows a Hammett $\sigma\rho$ correlation with a $\rho$-value of 4.0, a fairly large value indicative of significant negative charge close to the aromatic ring at the transition state.[103] Comparison of the rate of racemization of optically active ethylbenzene-$\alpha$-$d$ with the rate of deuterium exchange showed that replacement of hydrogen by hydrogen occurs with 82% net retention of configuration. These results indicated an ion-pair reaction mechanism (Scheme VIII).

The toluene substituent effect work was done by one of my early postdoctoral associates, Heinz Koch. Heinz was trained in classical organofluorine chemistry at Cornell but learned about isotopes and kinetics at Berkeley. We have shared a close personal friendship for many years. As a chemistry professor at Ithaca College he has had an outstanding teaching and research career entirely with undergraduates, ably assisted by his wife, Judy. His

Scheme VIII.

*Heinz Koch in Grimentz, Switzerland, in 1972, just after the first IUPAC Conference on Physical Organic Chemistry in Crans. Heinz was an early postdoctoral with me, and we have been personal friends for many years. His wife and two children all have some chemical connection to me. Heinz has been a Professor at Ithaca College for many years.*

# A Lifetime of Synergy with Theory and Experiment

*Professor Taniguchi, a prominent Japanese physical organic chemist, with Heinz Koch (right) during the Fifth Kyushi International Symposium on Physical Organic Chemistry (KISPOC) at Kyushi University in Fukuoka in 1993. Professor Taniguchi is one of the organizers of these symposia. I gave plenary talks at KISPOC I and KISPOC V.*

research has involved kinetics and isotope effect studies with highly halogenated compounds, a combination of his Ph.D. and postdoctoral traditions. Their son, Andy, recently obtained his Ph.D. at Berkeley under my direction, a rare case of father–son collaborators with me. Indeed, I now am connected to the entire family. Judy developed the study guide to the fourth edition of my coauthored textbook of organic chemistry (*see* section titled Textbook: *Introduction to Organic Chemistry*), and their daughter, Nanci, married Andy's labmate, Drew Speer, who also completed his Ph.D. with me. While they were undergraduates at Ithaca College, Nanci and Andy contributed to a research project involving the whole family and written up in a paper by Koch, Koch, Koch, and Koch, a paper now known as "Koch$^4$".[104]

Because of the benzyl anion character of the transition state, the reaction served the purpose for applying simple MO theory to the corresponding exchange reactions of α-deuterated polycyclic methylarenes. The experimental technique developed by Van Sickle was extended to these methylarenes by another early student, Bill Langworthy, who is now Vice-President for Academic Affairs at Fort Lewis College, Colorado. Comparison of the

Bill Langworthy, one of my early graduate students in carbon acidity, in his Berkeley lab in August, 1961. Bill was on the faculties of several colleges and universities before he assumed his present position as Vice-President for Academic Affairs of Fort Lewis College.

*From left: Dale Van Sickle, Bill Langworthy, Heinz Koch, and John Brauman, at a party at the Koch's in 1960. These four were early members of the carbon acidity group.*

relative rates with differences in Hückel π-energies (ΔM) showed the same type of dispersion into classes of compounds as shown earlier in carbonium ion reactions; α-naphthyl-type positions are less reactive than simple Hückel theory predicts (Figure 5).[105,106] The reactivity of 2-methylpyrene was an especial problem; Hückel theory predicts this compound to be less reactive than toluene, but experiment showed it to be 15 times as reactive. The 2-methylpyrene data point is on the far left of Figure 5, far from any correlation line. As the triangles in Figure 5 show, fluoranthene derivatives can also deviate seriously from simple correlations. Such experiments emphasized the limited quantitative usefulness of Hückel theory and forced our use of higher level theories. My further adventures in quantum chemistry are discussed subsequently (*see* section titled Ab Initio Quantum Organic Chemistry).

The initial impetus of my work in carbanion chemistry thus derived from the interrelation between theory and quantitative experiment; indeed, this early work was supported by the Air Force Office of Scientific Research as part of my general program

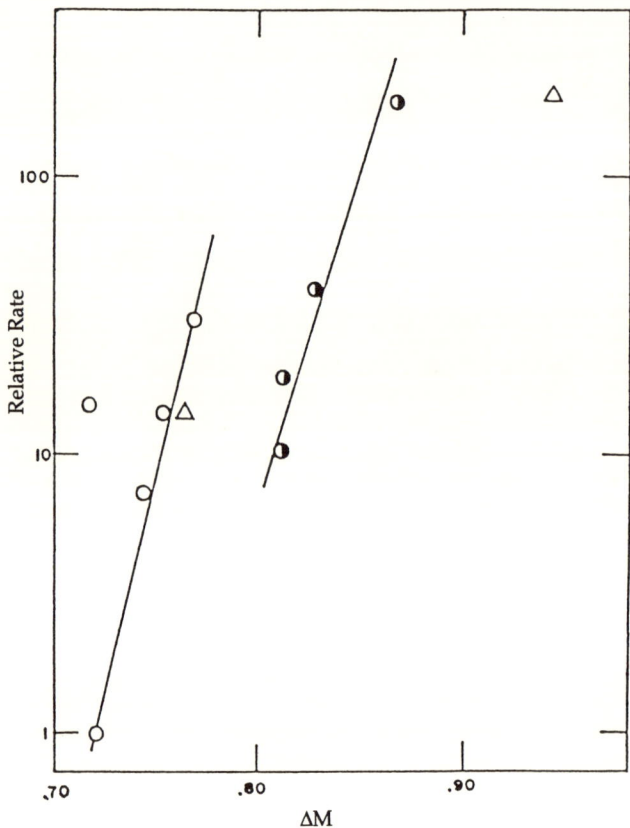

*Figure 5. Comparison of exchange rates of methylarene-α-d with lithium cyclohexylamide with Hückel molecular orbital calculations. Open circles are toluene-type methyls (the circle farthest left and far from the line is 2-pyrenylmethyl); half-filled circles are α-naphthyl-type methyls; triangles are fluoranthene compounds. ΔM is the calculated energy difference between the arylmethyl cation and the arene in units of the Hückel parameter β. (Reproduced from reference 106. Copyright 1963 American Chemical Society.)*

on experimental and theoretical studies of MO theory in organic chemistry. Nevertheless, I greatly enjoyed the experimental approaches involved: the use of deuterium and tritium isotopes, the development of inert atmosphere techniques, the analysis of complex kinetic problems, and the development and testing of new reaction mechanisms. This was fun! This fun aspect of chemistry never appears in papers or research proposals, but we all know (or should know) that the fun part of chemistry is

# A Lifetime of Synergy with Theory and Experiment

important to us who do it. Moreover, having gotten into carbanion chemistry, I learned that this field is rich and important, with significant questions of chemistry far beyond just the potential applications of simple MO theory. The lure of these additional questions concerning mechanisms of proton transfer, internal return, the stabilization and chemistry of ion pairs, the structures

*Playing ball at a group picnic in the 1960s. For many years the research group organized an annual picnic at one of the local regional park facilities.*

of organoalkali compounds, and the role of aggregation and the promise of exciting experiments proved irresistible. Thus, the original series of 50 research papers on "The Acidity of Hydrocarbons" was expanded with paper 51 to "Carbon Acidity".

The kinetic studies were extended by Ronald Lawler in the early 1960s to the acidities of aromatic ring protons. The different hydrogens in, for example, benzene, naphthalene, and pyrene showed a 50-fold range of reactivities, but a simple theory based on inductive effects was able to account for these relative reactivities.[107,108] For some of the reactive polycyclic hydrocarbons, colors developed during reaction that were indicative of the formation of radical anions. This hypothesis was confirmed by electron-spin resonance (ESR) studies. Ron subsequently became quite expert in ESR, and it has formed an important part of his independent research as Professor at Brown University. The results with biphenylene were combined with a number of other results with the generalization, "aryl positions adjacent to a fused strained ring have enhanced acidity and reduced reactivity toward electrophilic substitution."[109] This generalization was rationalized with a simple model based on the effective electronegativity of s-orbitals in atomic hybrids.

The exchange reactions of benzene with LiCHA are inconveniently slow (a kinetic run typically took a few weeks), and we looked for a more reactive base catalyst. Preliminary experiments in the early 1960s showed that CsCHA gives rates that are orders of magnitude faster, and we undertook a thorough study of this compound. CsCHA could be prepared by direct reaction of the amine with cesium metal. Cesium metal must be handled with great care. The metal has a low melting point, and it enflames on exposure to air. We tried different sources of the metal. Some pure samples were obtained commercially, but much of our cesium has been prepared by pyrolysis of the azide and triply distilling the product. Alkali azides are generally much safer than alkyl azides or heavy metal azides. They decompose on heating to give nitrogen and the alkali metal. On heating, cesium azide first melts and then decomposes rapidly (but not explosively). The rapid "boiling" and evolution of nitrogen from a black liquid followed by distillation of the golden-silver cesium makes for a rather spectacular experiment. The greater kinetic reactivity of cesium salts compared to lithium is generally attributed to the greater ionic radius of cesium cation; electrostatic bonding within an ion pair is reduced, and it is easier to separate charges even more as required in many reactions. Moreover, CsCHA is primarily monomeric in cyclohexylamine solution; LiCHA is highly aggregated, and the reactive monomers are present to only a small extent at typical reaction concentrations.

CsCHA is such a sufficiently strong base that later in the 1960s we were able to measure the hydrogen isotope exchange reactions even of saturated hydrocarbons. The reactions are slow; for example, cyclohexane is $10^{-8}$ as reactive as benzene.[110] In a typical experiment the half-life for reaction of cyclohexane was about 100 years. The measurements were possible because of the isotopic tracer nature of tritium. We used tritiated cyclohexylamine and followed the rate of incorporation of tritium into the hydrocarbon (eq 14):

$$C_6H_{12} + C_6H_{11}NHT \xrightarrow{CsCHA} C_6H_{11}T + C_6H_{11}NH_2 \qquad (14)$$

In this way we could follow the reaction to only 0.01–0.1% completion. Despite this very slow rate, two first-rate students, William R. Young and Richard A. Caldwell, working at different times and with somewhat different techniques, got excellent experimental agreement. By using perdeuteriocyclohexane, $C_6D_{12}$, we found that the reaction has a normal large isotope effect indicating that cyclohexylcesium is a reaction intermediate that does not react with solvent at diffusion-controlled rates. Other cycloalkanes showed the rate progression cyclopropane ($7 \times 10^4$) > cyclobutane (28) > cyclopentane (5.7) > cyclohexane (1) in quantitative agreement with estimates of the amount of s-character in the carbon orbital to hydrogen.[111] Figure 6 shows such a correlation for the cycloalkanes using the $J(^{13}C-^1H)$ coupling constant as a measure of the s-character in the C–H bond orbital. The straight line shows the excellent correlation for this group of hydrocarbons.

Dick Caldwell and Bill Young have gone on to successful careers of rather different types. Dick is a Professor at the University of Texas, Arlington, and Bill is a financial analyst for the chemical industry.

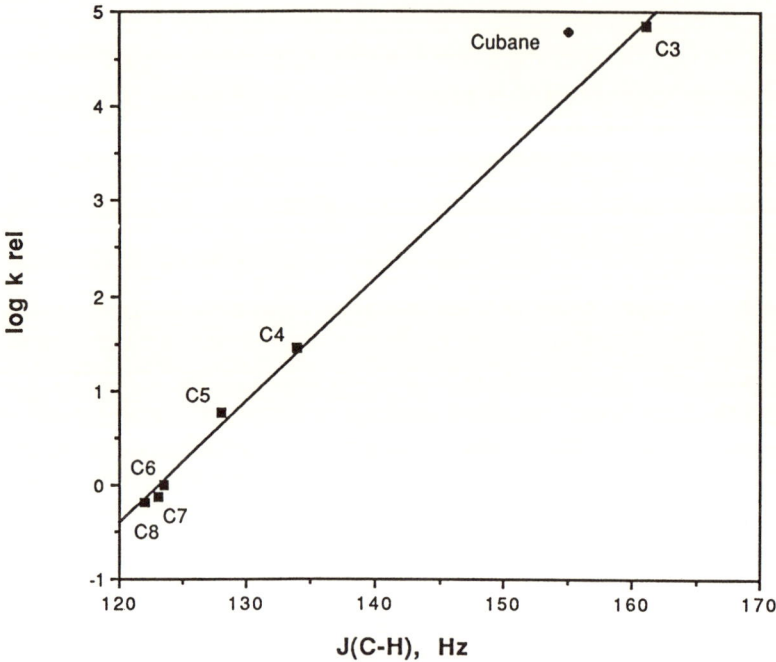

*Figure 6. Correlation of relative tritium exchange rates of cycloalkanes with CsCHA at 50 °C with J($^{13}$C–H) coupling constants. The corresponding value for cubane does not fit the correlation. (Reproduced from reference 112. Copyright 1991 American Chemical Society.)*

Methane was found to be more than 2000 times as reactive as cyclohexane but still much less reactive (relative rate = 0.03) than cyclopropane.[113] Ethylene, by comparison, is 0.08 as reactive as benzene or about 100 times as reactive as cyclopropane.[114] Recently, we have returned to measuring such proton-exchange reactions of hydrocarbons because of their potential relevance as a standard type of carbanion system for evaluating mechanisms of "hydrocarbon activation" reactions by organometallic reagents. The methyl hydrogens of 1,1,1-triphenylethane, $Ph_3CCH_3$, were found to exchange at a convenient rate, about midway between benzene and cyclohexane.[115] The primary isotope effect was determined by comparing the exchange rates of $Ph_3CCH_2D$ and $Ph_3CCH_2T$, and the secondary isotope effect was measured by comparing $Ph_3CCD_3$ and $Ph_3CCH_2D$. These studies showed that the transition state was similar to that of cyclohexane exchange even though the reactivities differ by several orders of magnitude.

In this work we used an approach to measuring tritium exchange that we developed earlier for toluene in water.[116] A common assay for tritium is by liquid scintillation counting. This method can give an accurate measure of the radioactivity content of a sample, but for a molar activity one needs to know how much material is in the scintillation vial. The determination of this amount is frequently difficult, particularly when small amounts of liquid substrates are involved. The method we developed was to use a sample doubly labeled with tritium and $^{14}$C. It is straightforward to determine both radioactivities in the same sample with modern scintillation counters and, because the $^{14}$C remains constant, the $^{14}$C activity measures the quantity of sample being counted, and the ratio $^{3}$H/$^{14}$C is the kinetic quantity that varies with time.

For some types of systems we have recently found[117] it convenient to use tritium NMR for studying these kinetics. Tritium is a somewhat more sensitive NMR nucleus than $^{1}$H, and only a few parts per million of tritium in hydrogen will give a good NMR signal. This approach was applied to the kinetic acidity of cubane; as shown in Figure 6 its hydrogens are somewhat more reactive than expected for the amount of s-character in the C–H bond as measured by the $^{13}$C–H coupling constant.[112] The significance of this result must await further work with related polycyclic compounds; that is, the coupling constant as a property of the hydrocarbon is clearly not a sufficient measure of the nature of the proton transfer transition state. With additional experimental work we plan to explore other properties that may be better indicators of reactivity, such as the computed charge of the proton.

## Equilibrium Acidities in Cyclohexylamine

More acidic hydrocarbons of the type of fluorene and di- and triarylmethanes are deprotonated completely by CsCHA. For these compounds with highly delocalized carbanions it was possible to measure equilibrium constants for the ion-pair equilibria (eq 15).

$$R^{-} Cs^{+} + R'H \underset{}{\overset{K}{\rightleftharpoons}} RH + R'^{-} Cs^{+} \quad\quad (15)$$

*Giving a seminar on my early carbon acidity research at the Universal Oil Products Company in 1963.*

## A Lifetime of Synergy with Theory and Experiment

The cesium salts in eq 15 have absorption spectra in the visible and near-UV that could be determined independently. The spectrum of the equilibrium mixture and stoichiometry then give the equilibrium constant, $K$. The logarithms of these equilibrium constants are p$K$ differences between the two hydrocarbons on an ion-pair acidity scale. This method is a quantitative version of an approach used by McEwen in 1936.[118] McEwen used equilibria of organosodium compounds in benzene but with the methods available at that time could obtain only rough measures of the equilibrium constants.

The equilibria in eq 15 define ion-pair acidity differences between compounds. Chemists, however, are more used to thinking with absolute p$K$ numbers, usually for aqueous solutions. Hydrocarbons are not sufficiently acidic and soluble in water for their acidities to be measured in a normal manner. Hammett[119,120], however, showed how reasonable assumptions with regard to ratios of activity coefficients would permit the derivation of aqueous p$K$ values from measurements in partially aqueous and nonaqueous solutions. This method, now known as the Hammett acidity function method, has been applied[121] to derive p$K$ values for a number of weak acids and bases. By the early 1960s, this method had been applied[122,123] to a number of relatively acidic hydrocarbons in partially aqueous solutions. These compounds, with p$K$ values in the low twenties or lower, all have one or more cyclopentadiene moieties in their structures. At this time, Kuhn[124–127], for example, showed how carbanions with two or more conjugated fluorenyl groups could give hydrocarbon p$K$ values of 10 or less.

In order to put our measured p$K$ differences on an absolute p$K$ scale, we chose one compound, 9-phenylfluorene, as an arbitrary standard and set its p$K$ at the value, 18.5, derived earlier from acidity function measurements in other solvents.[122] In this way we obtained[128,129] by the mid-1960s a series of p$K$ values for a number of aromatic hydrocarbons. These were among the first quantitative equilibrium acidities for hydrocarbons of such low acidity. Additional p$K$ values of substituted fluorene compounds by the acidity function method appeared soon thereafter.[130,131] Within the limits of the acidity function approximation, these p$K$ values are roughly those of an aqueous scale and agree well with the ion pair values in cyclohexylamine.

Meanwhile, a number of workers[132-137] were determining acidities in various aqueous and other mixtures of dimethyl sulfoxide (DMSO) using the acidity function approach. Ritchie and Uschold[138-140] applied potentiometric titrations in DMSO to obtain pK values for the DMSO standard state. This work was extended by Fred Bordwell°, whose research group has amassed a large number of p$K_a$ values in DMSO.[141,142] DMSO is a polar solvent, and delocalized carbanions of the fluorenyl and triarylmethyl types are completely dissociated in this solvent. Nevertheless, the relative pK values of the ion-pair acidities of the cesium salts in cyclohexylamine give an excellent linear correlation with the ionic pK values in DMSO (Figure 7), even though one set refers approximately to an aqueous standard state and the other refers to the DMSO standard state. The correlation with a slope close to unity shows that for highly delocalized carbanions, in which the negative charge is spread over a large volume, solvation energies and ion-pairing energies with a large cation do not vary much from one structure to another.

The close comparison of the acidities of conjugated hydrocarbons in different solvents was affirmed[143] by studies of highly acidic hydrocarbons in cyclohexylamine itself. Some compounds of this type are sufficiently acidic that they form cyclohexylammonium ion-pair salts of carbanions to some extent. By estimating the dissociation constants of the ion pairs we could derive the actual ionic p$K_a$ values of these compounds in cyclohexylamine. We found that these p$K_a$ values are approximately the same as the corresponding aqueous pK values[144]; that is, the higher basicity of cyclohexylamine compared to water approximately compensates for its lower polarity so that pK values with the pure solvents as standard states are about the same, at least for such delocalized carbanions. Thus, the derived CsCHA "ion-pair pK values" that were shown already to approximate the ionic pK values in aqueous solution also approximate the absolute ionic pK values in cyclohexylamine!

These hydrocarbons have conjugate bases that are highly delocalized carbanions and form a valuable set of reference

---

° See Appendix A for brief biographical information about each person designated with the ° symbol.

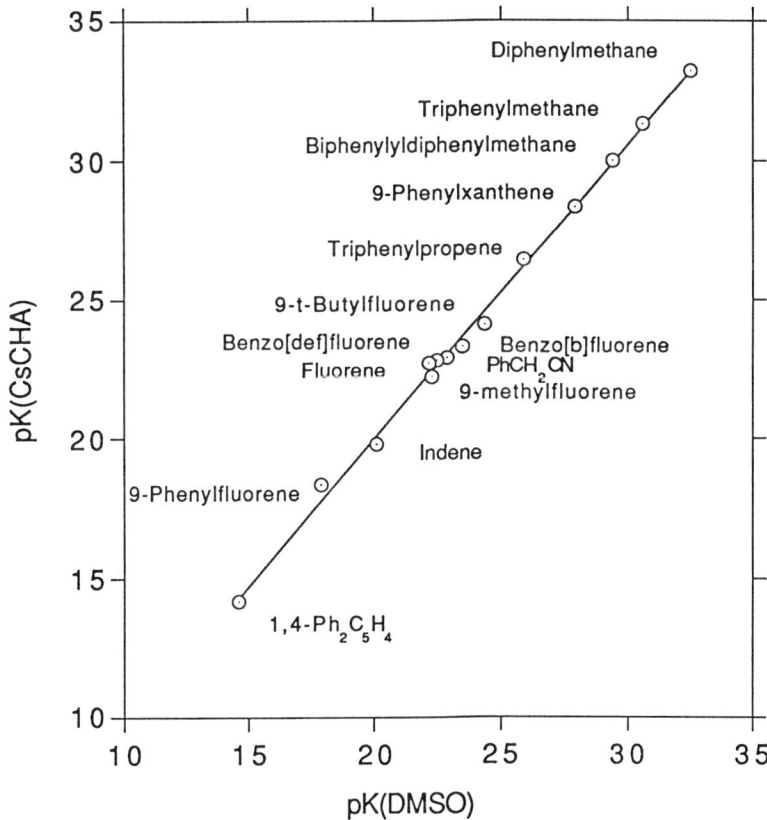

*Figure 7. Relative ion-pair pK values of cesium salts in cyclohexylamine compared with ion pK values in dimethyl sulfoxide (DMSO). The correlation line is pK(CsCHA) = −1.123 + (1.056)pK(DMSO); r = 0.998.*

indicators for finding specific effects in other systems. For example, phenylacetylene forms a localized carbanion in which the electrostatic interaction to a cation is greater. Accordingly, in ion-pair equilibria phenylacetylene has been found[142–145] to be more than 5 p$K$ units more acidic than in DMSO in which free ions are involved.

The localized charge of the phenylacetylide ion is bound electrostatically to the small lithium cation relatively more firmly than the reference delocalized carbanions would suggest. Many of these results on equilibrium ion-pair acidities involving localized and delocalized carbanions in cyclohexylamine have been reviewed.[144, 146]

## Brønsted Correlations

Important differences between localized and delocalized carbanions were also discovered in studies relating kinetic and equilibrium acidities. The comparison of rate and equilibrium often takes the form of a Brønsted correlation (eq 16):

$$\log (k/k_0) = \beta \log (K/K_0) \tag{16}$$

A series of polyarylmethanes, such as di- and triphenylmethane and $p$-methylbiphenyl, provided a suitable test in research that started in the 1960s and has continued to the present. We were able to measure hydrogen isotope exchange rates in the benzylic position with LiCHA just as in the parent case (i.e., toluene) already discussed. For many of these same compounds we were also able to measure relative equilibrium ion-pair acidities for the cesium salts in cyclohexylamine. Comparison of the two sets of measurements gives the linear Brønsted plot in Figure 8.[147] The slope of this correlation, 0.31, is typical of such Brønsted comparisons of exchange rates and equilibria for delocalized carbanions. For these cases, the slope $\beta$ undoubtedly is a valid measure of the degree to which the reaction has progressed at the transition state. That is, the transition state undoubtedly has a partially pyramidal central carbon with a substantial negative charge; conjugation to this carbon and delocalization of the charge is only a fraction of what it is in the product carbanion. This result is different from the effect of polar substituents on the exchange rates of toluene discussed previously. The Hammett $\sigma\rho$ correlation (which gave $\rho$ = 4.0) gives an indication of the amount of charge at the $\alpha$-position, but conjugating groups, especially nonpolar ones such as hydrocarbon $\pi$-systems, are better for showing the effects of charge *conjugation* at the transition state.

Although toluene is too weakly acidic to measure directly with CsCHA, the short extrapolation to the rate of exchange of toluene with LiCHA gives a p$K$ value of 41 (per hydrogen) on the CsCHA scale.[148] The determination of an accurate value for this fundamentally important compound had long been a goal in my research group. The number so derived is much greater than the approximate value of 35.5 on Cram's MSAD scale based on cruder

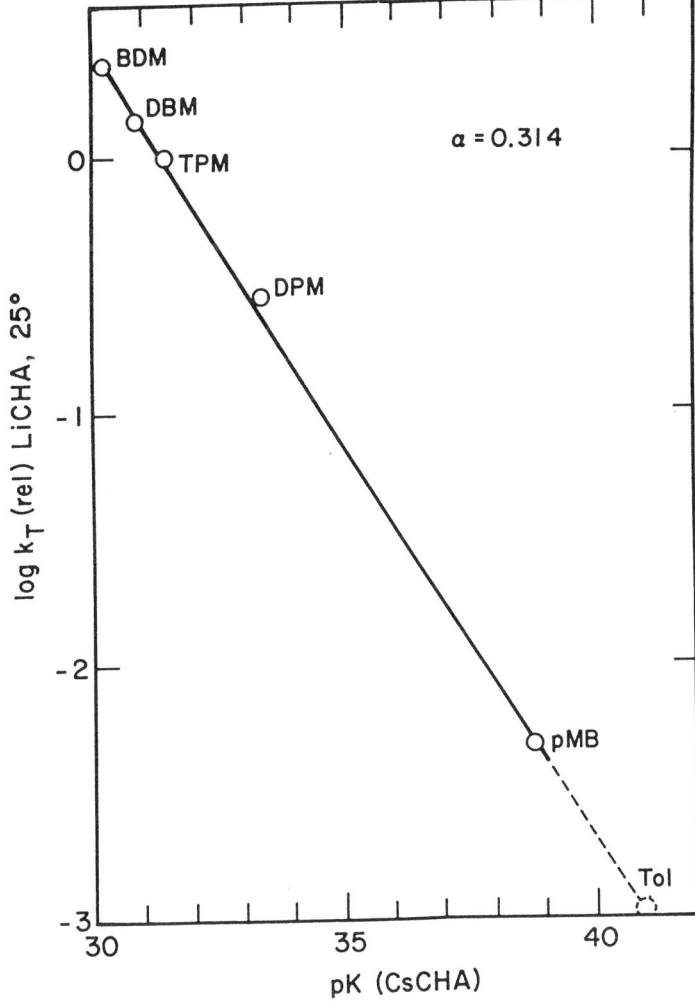

*Figure 8. Brønsted correlation of arylmethanes between tritium exchange rates with LiCHA and relative pK values based on equilibria of cesium salts in cyclohexylamine. Abbreviations: DBM is p-biphenylyldiphenylmethane; DBM is di-p-biphenylylmethane; TPM is triphenylmethane; DPM is diphenylmethane; pMB is p-methylbiphenyl; Tol is toluene. The dashed extrapolation gives a pK value for toluene of 41. (Reproduced from reference 147. Copyright 1973 American Chemical Society.)*

measurements of earlier decades (MSAD = McEwen–Streitwieser–Applequist–Dessy).[149] More recently (1985), we obtained a pK value of toluene of 40 in water. We made use of the slight but significant solubility of toluene in water and measured the rates of tritium exchange in aqueous sodium hydroxide at high temperatures in a Monel bomb. These rates were extrapolated to room temperature and combined with the known rate of reaction of benzyl anion with water (derived from radiolysis experiments)[150] to deduce the equilibrium constant.[116]

These experiments were done by Jiu Xiang Ni, whom I met during a trip to China in 1981. Ni's career as a chemist was interrupted by the Cultural Revolution, and his chemistry showed important gaps. I was impressed by what he had been able to accomplish under trying circumstances and later was able to arrange a stay in my laboratory where he did this and other experimental work. He subsequently went on to graduate studies and received his Ph.D. degree from the University of California, Santa Cruz. His daughter, Hai-Ye Ni, is an outstanding cellist with a growing international reputation.

These hydrocarbons that are so useful as indicators are sufficiently acidic to also undergo hydrogen isotope exchange reactions with methanolic sodium methoxide at reasonable rates. For example, the reaction of fluorene-9-$t$ with 0.1 M NaOCH$_3$ has a half-life of 5 hours at 45 °C and a normal primary isotope effect.[151] A number of fluorene-type hydrocarbons (e.g., benzofluorenes, indene, and phenylfluorenes) give an excellent Brønsted correlation with the corresponding cesium ion-pair acidities, with a slope β = 0.37 (Figure 9), a typical value in such correlations; this correlation indicates, as in the previous case of arylmethanes, the degree to which conjugation at the transition state has proceeded compared to product.[151,152] The kinetic measurements were straightforward because no special care is required to exclude oxygen or moisture; thus, this work was excellent training for undergraduate students or for introducing new students to kinetic measurements and the use of isotopes.

A similar study was carried out with arylmethanes. These compounds are much less reactive; for example, the exchange reaction of triphenylmethane-$t$ with 0.1 M NaOCH$_3$ at 45 °C has a half-life of 350 years. The kinetic measurements were run at higher temperatures and extrapolated to 45 °C. One interesting aspect of these reactions is that the Brønsted correlation of these arylmethanes (such as di- and triphenylmethane and $p$-biphenylyldiphenylmethane) has a slope of 0.58; that is, this Brønsted

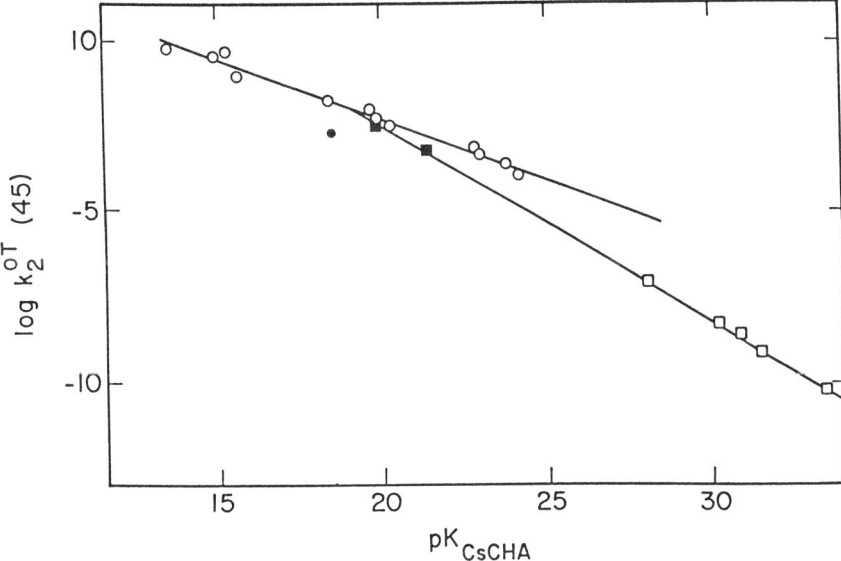

*Figure 9. Brønsted correlations of tritium exchange rates of indicator hydrocarbons in methanolic sodium methoxide with CsCHA pK values. Symbols: open circles are fluorene, indene, and cyclopentadiene compounds with a Brønsted slope of 0.37; open squares are arylmethanes with a slope of 0.58; filled squares are the phenalene derivatives (benzo[cd]pyrene, 7, and benzanthrene, 8) that fit the arylmethane correlation; filled circle is phenalene, 5, which is all by itself.*

correlation is different from that given by the fluorenes (Figure 9).[153] This finding that hydrocarbons containing the cyclopentadienyl skeleton form a different Brønsted family from the arylmethanes was totally unexpected. A number of previous Brønsted-type studies comparing rates and equilibria had incorporated fluorene and di- and triphenylmethane within the same correlation by assuming that these hydrocarbons are part of the same Brønsted family. Our work demonstrates that "closely related" compounds need to be studied carefully to ensure that they are indeed "closely related"! We recently added some derivatives of phenalene, 5, to this study. Deprotonation of 5 gives the phenalenyl anion for which six equivalent resonance structures, 6, can be written. Benzo[cd]pyrene, 7, and benzanthrene, 8, are the filled squares in Figure 9 and clearly behave as arylmethane-type derivatives. The parent compound, phenalene (filled circle in

[Structures 5, 6, 7, 8]

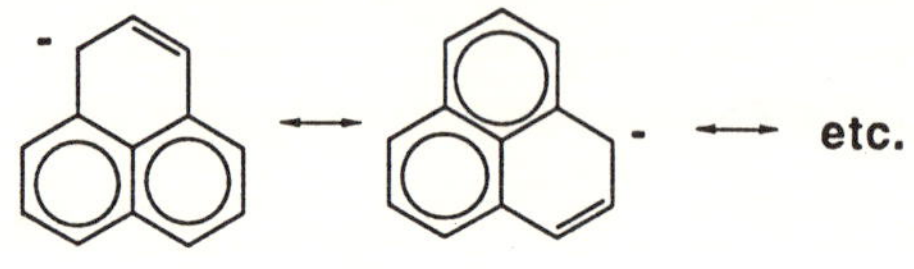

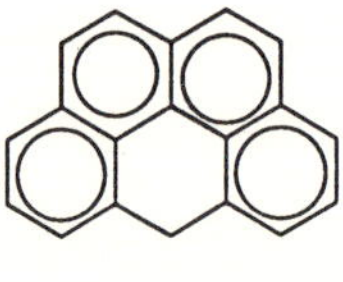

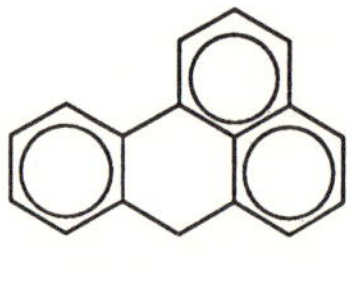

Figure 9), does not fit either correlation and is apparently a member of a still different Brønsted family![154]

Some exchange rates were also measured[155] for some 2-substituted fluorenes with methanolic sodium methoxide. Comparison with equilibrium acidities gives a Brønsted slope of 0.66, higher than that found for related compounds without polar substituents (Figure 9). This difference is related to the then unexpected finding of Bordwell[156,157] and Schechter[158] of Brønsted coefficients greater than unity for acidities and exchange reactions of arylnitroalkanes. Because the exchange transition state lies between reactants and products on a reaction coordinate diagram it had normally been assumed that Brønsted slopes must lie between 0 and 1. Schechter and Bordwell's unusual result was subsequently interpreted by Kresge.[159,160] The equivalent explanation of the present results makes use of the transition-state structure 9.

In the equilibrium anion the charge is distributed about both rings and is stabilized by inductive interaction with the substitu-

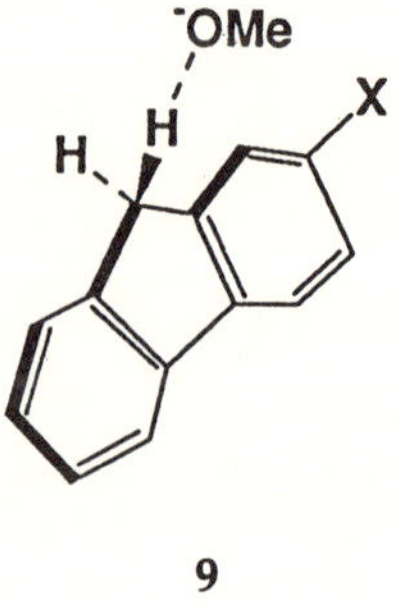

[Structure 9]

ent dipole. In the proton-transfer transition state the charge is concentrated at the methoxide oxygen and at the still pyramidal central carbon. The reduced delocalization of the charge at the transition state is indicated by the Brønsted slope of 0.37 in Figure 9. The polar substituent in **9**, however, is effectively closer to the charge in the transition state than in the delocalized carbanion and produces a larger Brønsted slope. This comparison also points up the use indicated previously of conjugating nonpolar substituents to tell something different about transition states compared to the use of polar substituent effects.

These different types of results illustrate why I have found the topic of carbon acidity so interesting. The understanding of acidity and basicity lies at the heart of physical organic chemistry and involves both thermodynamics and kinetics. Our work has extended the earlier studies of aqueous acids and bases and ions to ion pairs of much less acidic compounds. My involvement in this subject has continued to the present day.

## Consulting at DuPont

One of the interesting applications of such experimental measurements of carbon acidities is to the question of fluorine anionic hyperconjugation. This concept, which can be expressed as a significant contribution of the "no-bond" resonance structure in β-fluoroethyl anion (Figure 10), was proposed by Roberts[161] in 1950 to interpret some substituent effects of the $CF_3$ group and is derived from an earlier interpretation of bond shortening in carbon tetrachloride by Pauling.[162]

I had become interested in organofluorine chemistry through my consulting at the DuPont Company. Early in 1957, I had been invited to a roundtable discussion at the Polychemicals Department at the DuPont Experimental Station in Wilmington, Delaware. I was glad to have this opportunity to learn about some industrial chemistry since I had gotten my Berkeley offer so early in my postdoctoral year at MIT that I never had done any industrial interviewing. The discussion group included several other young assistant professors from different universities, and we were presented with some of the interesting chemistry done at

$H_2\overset{\ominus}{C} - CH_2 \curvearrowright X \quad \longleftrightarrow \quad H_2C = CH_2 \quad X^-$

*Figure 10. Anionic hyperconjugation or "double-bond–no-bond" resonance in β-fluoroethyl anion.*

DuPont. I greatly enjoyed this roundtable. I was quite relaxed and natural during these discussions, never realizing in my youthful naiveté that we were being looked over as potential consultants. I was therefore totally surprised when I was offered a consulting contract several weeks later, and I have been a consultant at DuPont ever since.

There is an amusing irony in all of this. Because I lived in Long Island during the time of the 1939–1940 New York World's Fair, I was able to go to the fair by myself on several occasions. I used to particularly enjoy the DuPont exhibit, and I can still recall the demonstrations of Lucite, synthetic fibers, and polymerizations. To become a DuPont chemist then became my dream. Little did I realize that the dream would be realized but not in quite the way I had imagined.

My consulting trips usually started with a discussion group consisting of Frank Gresham°, Director of Pioneering Research of the then Polychemicals Department, and several group supervisors. I had met Frank at the roundtable and had frequent contact with him for a number of years. He was an astute and effective scientist and had an important effect on my own chemistry. In these group meetings we often discussed unusual chemical results that frequently defied rational interpretation by the conventional wisdom of the day. The Polychemicals Department was a well-run and productive department that provided many of the leaders of the DuPont Company in more recent years. Many of these leaders started their DuPont careers in Frank Gresham's group.[163] Research in his group included fluorocarbon chemistry of novel monomers and polymers and introduced me to the fascinating field of organofluorine chemistry. Thinking about this chemistry had an inevitable and positive effect on my own research; this thinking led naturally to experimental and theoretical studies of fluorinated carbanions. In the summer of 1965 I attended the Third International Fluorine Symposium in Munich. This was only the second European meeting I had been able to attend. The reason requires a digression into personal history and its important effect on my science.

## Mary Ann and Suzanne

I met Mary Ann Good at a party while in graduate school. We fell in love and married in the summer of 1950 on the strength of a graduate research fellowship I was granted that had a stipend of $1200 a year. Our first child, David, was born in 1953, the year following our move to Berkeley. Our daughter, Susan, was born in February of 1956. During this Berkeley period, Mary Ann started having bouts of depression that became increasingly severe. Despite some initial psychiatric consultations that fall, she attempted suicide and required hospitalization. After a period of hospitalization in psychiatric wards of county and private hospitals, she was admitted to the Langley Porter Clinic, a well-regarded teaching institution associated with the University of California Medical School in San Francisco. She was treated there for several months. Meanwhile, I had two children to care for and school responsibilities. Susan was placed in a foster home and David was just old enough to be accepted by the Berkeley Day Care Center. I was able to take him there in the morning and pick him up after a day's work at the University. It's a pity that this great institution no longer exists. These social services available almost four decades ago exceed those of today, and I wonder what a modern Assistant Professor would be able to do if put into similar circumstances.

Mary Ann was given weekend furloughs and became an outpatient after several months. But it was a year and a half before we were able to bring our daughter back home and become a united family once again. Any story involving the return of a child still has a strong emotional impact on me, many decades after the event. My salary at the time was quite low, and DuPont's consulting offer in 1957 came at an important time. Langley Porter charged very little, but they eventually discharged Mary Ann and she continued private psychotherapy for many years. This treatment, which was not covered by health insurance, helped to

*With my first wife, Mary Ann, in the apartment of Jack and Edith Roberts in early 1952, during my postdoctoral year at MIT. (Photo taken by Jack Roberts.)*

# A Lifetime of Synergy with Theory and Experiment

*My first wife, Mary Ann, and me on the occasion of my first award, the California Section Award of the ACS, in 1964. This was one of her best periods in freedom from depression, but the depression returned not long thereafter.*

lessen her pain but did not by any means cure her. I believe that the underlying cause of her depression was a malfunction of brain chemistry and that psychotherapy can have only limited effectiveness in such a case. Mary Ann had a warm personality, and she had many friends. But she lived in an almost continuous cloud of depression that varied greatly in intensity. She was, nevertheless, able to mask its effects; in group situations, for example, only her closest friends and I were aware of her true state. She was also careful to point out to the children that they were in no way responsible for her periods of depression. Her depression occasionally was intense, and there was always the possibility of suicide. When I was away from her I feared the telephone. On one occasion, I had just started a consulting day at DuPont and was called home immediately because of a particularly severe bout.

The year 1964 was a good year, with her depression generally at a minimal level. We took a vacation drive that summer to southern California, taking the children to Disneyland and spending a pleasant week in La Jolla, where I gave a research talk at the University of California, San Diego. That fall I received the California Section Award of the local section of the American Chemical Society. One's first major award is always special—it is either the first award or the only award. Accordingly, Mary Ann arranged a celebration party with a number of colleagues and friends. They included Al Wolf, who spent the fall semester at Berkeley as a Visiting Professor teaching organic chemistry.

But Mary Ann's freedom from depression didn't last. In the spring of 1965 she had to return to psychiatric care. By this time the first antidepressant drugs had appeared and her psychiatrist prescribed one for her. She took the first of this prescription on the morning of May 12, and it had the opposite effect from that intended. That afternoon I was in my office discussing research with several students when I received a phone call from a California Highway Patrol officer telling me that my wife had jumped to her death from the Oakland Bay Bridge.

I had lived with this possibility for almost nine years, but the event was still sudden and unexpected. We had planned a family outing the following weekend at an inn in Inverness, a quiet resort town on Tomales Bay close to the ocean and not far from Berkeley. I made the trip anyway with the children to emphasize that we were still a family even without a mother. We were accompanied by our good friend, Frank Goodrich, and his two children, who are the same ages as my children. I had known Frank slightly at Columbia where he was a graduate student in physical chemistry. He took a position at the then California Research Corporation (now Chevron Research) in Richmond, California, in 1952, and thus he and his wife, Madeleine, and we were newcomers to California at the same time and with a shared background. Our friendship grew as both women became pregnant twice at the same times, and our children were christened together in the same Unitarian Church. Frank played a major role in organizing Mary Ann's memorial service at this church. The Goodriches left California that summer for Potsdam, New York, where Frank started his own academic career at Clarkson Institute of Technology. He died prematurely in 1980.

The period of bereavement following Mary Ann's death included times of intense loneliness in which I was enormously

## A Lifetime of Synergy with Theory and Experiment

comforted by having my two children with me; they provided the emotional support of being part of a family. The effect on my chemical output is a matter of record. My substantial publication list in 1965 includes books and papers completed before Mary Ann's death. For 1966, my bibliography is almost devoid of entries, but 1967 saw a rebound with 10 publications.

I was fortunate not to have the additional complication of classroom teaching responsibilities at the time of Mary Ann's death. I had been appointed a Miller Research Professor for 1964–1965, a position that carries with it independence from classroom teaching and administration. The Adolph C. and Mary Sprague Miller Institute for Basic Research in Science was established by the Regents of the University of California in 1955 with a bequest from the Millers. It provides several prestigious postdoctoral fellowships in the sciences as well as research professorships at the Berkeley campus. At that time the faculty recipients in chemistry were expected to find their own temporary replacements. I was fortunate to have found two excellent teachers for my course, Al Wolf for the fall semester and Tom Katz[°] of Columbia University for the spring semester. In later years I served on the executive committee and as Director of the Miller Institute.

Mary Ann's death marks a division of my life into two rather different parts. During the year following my wife's death I dated again and relearned the games that men and women play in the courtship ritual at several levels. As a widower with children I was in a somewhat different category from most single men. I became aware of the different lives of the marrieds and the singles and how little these lives overlap. I was able to meet many different people, and I learned a great deal from them. It is a cliché that time heals, but all clichés become such because they embody truth. Mary Ann remained a vivid memory, but the pain faded.

This year ended with one of those "small world" stories that defy fiction when I met my present wife, Suzanne. A blind luncheon date was set up for May 10, 1966, by a mutual friend, a former neighbor of mine, Gloria Goldberg. Gloria's daughter, Heidi, and my daughter Susan are of an age and were school chums. Gloria is also the half sister of Ernest Grunwald,[°] who was

*My second wife, Sue, in 1966 before our engagement.*

a postdoctoral student with Professor Louis Hammett in 1949 and shared a laboratory with me. Suzanne was living at the time in Sausalito, a town across the bay on the other side of the Golden Gate Bridge. Her father was the Unitarian minister who christened both of my children and presided at my first wife's memorial service. I knew little about him at the time of this service and would not have imagined how closely our subsequent lives were to become intertwined. I had just started going to the San Francisco Opera in the fall, 1966, season when I discovered that *Boris Godounov* is one of Sue's favorite operas. (We call my wife "Sue" and daughter "Susan".) Sue had had little opportunity to see opera, but she used to listen to the Metropolitan Opera broadcasts regularly. Thus, my next date with her was at the San Francisco Opera's production of *Boris Godounov*. Together we have both developed our love for this "ultimate art"[164] and have since seen more than 400 opera performances.

Our relationship developed rapidly thereafter. We started "going steady" later that fall, became engaged on New Year's Eve, and were married by Sue's father in July, 1967. For our honeymoon we drove to Idaho with my two children, David and Susan, then 13 and 11, two of their friends, and a dog. Sue's parents at that time owned a summer cabin in Island Park, Idaho, prime flyfishing country near Yellowstone Park. Her father was an avid fisherman, and he taught David and me to love the sport. Although Sue's parents sold their cabin in the late 1970s, Sue and I obtained property of our own in Island Park and in recent years have spent several weeks there every summer. We are joined for one week by David, who takes this vacation from his wife and daughter so that father and son can go fishing together for a few days each year. Most fishing is stochastic fishing in which one waits for a random fish to bite. Some flyfishing is also like that, but the greatest thrill comes in dry-fly fishing when one sees a fish, fishes toward it, and catches it—catch-and-release with barbless hooks, of course.

Sue was a brave woman to enter an established family at the age of 33. Indeed, she adopted both children early in our marriage. She thus became their legal mother and not just a stepmother. Our family life together has been normal, and both children have entered adulthood successfully. David is an emergency room physician in Phoenix. Susan teaches voice and with her band, Susan's Room, she and her husband produce their own records (Zanna Discs) in Los Angeles. Whatever the part I played in their successful transition into adulthood, this result is surely my greatest accomplishment.

## Organofluorine Carbanions

Shortly after Mary Ann's death I was able to hire a reliable full-time housekeeper to whose care I was able to entrust the children. Thus, I was able to attend the fluorine conference in Munich. Munich is a great city and had a beneficial effect on my emotional state. I have since been often in Munich, and it is now my second favorite city in the world (San Francisco, naturally, is first). The meeting in Munich also had an important scientific effect for I met

My children, Susan and David, in March, 1977, taken at our Berkeley home, when they were 21 and 23 years old, respectively. Susan teaches voice and produces records (Zanna Discs) of her band, Susan's Room, in Los Angeles. David is an Emergency Physician at a hospital in Phoenix, AZ.

# A Lifetime of Synergy with Theory and Experiment

J. Colin Tatlow, Professor Emeritus, Birmingham University. I met Colin at a fluorine meeting in Munich in 1965, and he sent generous gifts of fluorinated bicyclic compounds with which we made important acidity measurements.

there a number of internationally renowned fluorine chemists. One was Professor J. C. Tatlow of Birmingham, England, who had just prepared the first of a series of highly fluorinated bicyclo[2.2.1]heptanes.[165-167] These important syntheses provided a crucial test for the role of anionic hyperconjugation. Professor Tatlow and his associate, Dr. R. Stephens, sent us generous samples of compounds **10** and the related bicyclo[2.2.2]octane compound **11**.

David Holtz was one of my graduate students at that time (1964–1968). Using these samples, he obtained rate constants for tritium exchange of the bridgehead hydrogens with methanolic sodium methoxide. The work was not straightforward because these compounds are exceptionally reactive, but Dave worked out clever methods for getting rate constants at low temperatures, down to –74 °C. As an example of the high reactivity of these

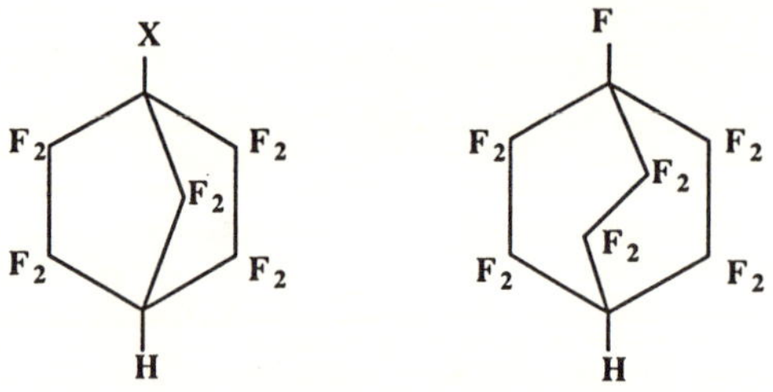

**10**  **11**

X = H, CH₃, F, CF₃, Br, I

compounds, **10** (X = F) undergoes complete tritium exchange in less than 4 minutes at –23 °C with 0.001 M NaOCH$_3$ in methanol containing CH$_3$OT.[168] This reactivity is of the same order of magnitude as that reported earlier by Andreades[169] for (CF$_3$)$_3$CH. This is the same Sam Andreades mentioned previously as one of my earliest graduate students. After synthetic and mechanism studies at Berkeley he went to DuPont and did pioneering work in organofluorine chemistry in the Central Research Department before transferring to the Photoproducts Department. His work with (CF$_3$)$_3$CH had to be done with extreme care because the elimination product, perfluoroisobutylene, (CF$_3$)$_2$C=CF$_2$, is a highly poisonous gas. He interpreted the high kinetic acidity of (CF$_3$)$_3$CH in terms of fluorine hyperconjugation. In the bridgehead carbanion from **10**, however, such hyperconjugation must be much less important because no-bond resonance now involves a bridgehead double bond, **12**; thus, Dave Holtz's result shows that such high kinetic acidity does not require anionic hyperconjugation.

The primary isotope effect, $k_H/k_D$, was found to be close to unity, a result indicative of internal return.[170] The importance of

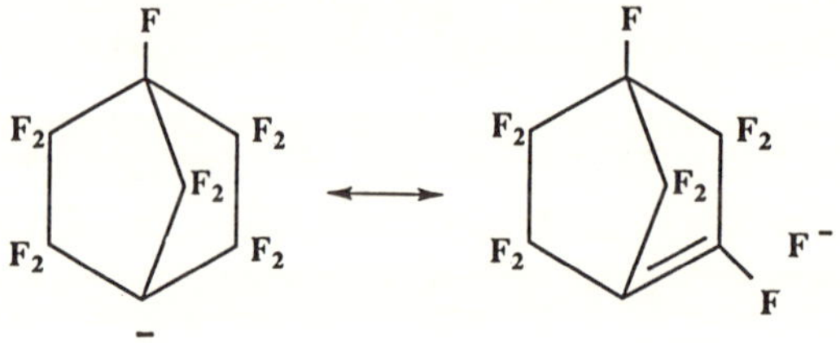

**12**

# A Lifetime of Synergy with Theory and Experiment

*For my 38th birthday in 1965 the research group presented me with a special present, a Gilbert chemistry set, perhaps to help improve my laboratory technique. My secretary, Lynne Gloria, had this set mounted and displayed in her office for a number of years. Members of the group shown (from left): Herschel Rabitz, Dave Holtz, Bob Bittman, and Mark Bixler. (Picture taken by S. Ozawa.)*

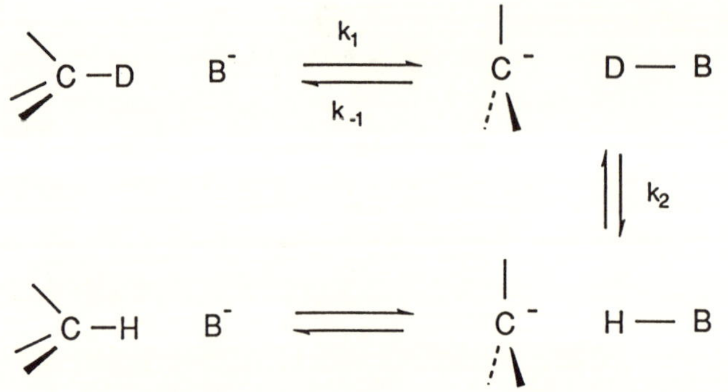

```
\C-D    B⁻   ⇌[k₁][k₋₁]   C⁻    D — B
/                         /\

                                ⇅ k₂

\C-H    B⁻   ⇌             C⁻    H — B
/                         /\
```

Figure 11. Cram's internal return mechanism.

$$k_{exp} = \frac{k_1 k_2}{k_{-1} + k_2} \quad (17)$$

internal return in carbanion reactions was first demonstrated by Cram[149,171] on the basis of stereochemical and isotope effect studies. The effect is illustrated by the kinetic analysis in Figure 11. The experimental rate constant, $k_{exp}$, is given in terms of the mechanistic rate constants, $k_1$, $k_{-1}$, and $k_2$, by eq 17. There is a fundamental distinction in this equation between $k_{exp}$, an empirical experimental parameter based on pseudo-first-order conditions, and the rate constants defined by a mechanistic hypothesis.

When the solvent is only weakly bonded to the intermediate carbanion, the replacement of the labeled solvent by bulk solvent is rapid and $k_2 \gg k_{-1}$; eq 17 then reduces to $k_{exp} = k_1$. That is, the normal experimental exchange rates are then those of the first step, and the isotope effects measured when D is replaced by T, for example, are also those of the first step. This condition holds true for exchange reactions of fluorene with methanolic sodium methoxide. If the intermediate is stabilized, however, such that $k_{-1} \gg k_2$ (important internal return), then eq 17 reduces to $k_{exp} = (k_1/k_{-1})k_2$; that is, the measured rate constant is now a composite, and the measured isotope effect is not simple. Nevertheless, neither the ratio $(k_1/k_{-1})$ nor the solvation rate $k_2$ is expected to have an isotope effect far from unity; thus, an experimental isotope effect of close to unity is indicative of extensive internal return. In practice, such internal return is found to be particularly important for localized carbanions for which hydrogen bonding phase and in DMSO.[174] The greater acidity of $CH(CF_3)_3$ was attributed primarily to anionic hyperconjugation, but here also the more acidic compound has more fluorines closer to the carbanionic center.

A second test of the concept of anionic hyperconjugation was provided by exchange rates of some 9-substituted fluorenes. Exchange rates for the substituents $CH_3$, $CH_2Ph$, $CH_2OCH_3$, and $CF_3$ give a good linear correlation with the acidities of the corresponding substituted acetic acids.[175] That is, all of the substituents fit a simple inductive effect correlation and no special enhancement was found for the $CF_3$ group. If anionic hyperconjugation had a dominating effect, we would have expected the fluorosubstituent to provide substantially greater reactivity than a simple inductive effect would suggest. Because the $CF_3$ group gives the reactivity expected for inductive effects alone, anionic hyperconjugation must be unimportant in this case.

These experiments would seem to settle the matter. Indeed, Dave Holtz reviewed a number of other properties of fluorinated compounds and concluded in two reviews written during his postdoctoral work at Cal Tech in 1968–1970 that anionic hyperconjugation is unimportant as a phenomenon in chemistry.[176] Dave's subsequent career took several unusual turns. After Cal Tech, he served for several years as a staff scientist on an environment committee of the National Academy of Sciences in Washington and later he turned to business, eventually earning an MBA. He now runs his own financial consulting business as a CPA in California.

Unfortunately, chemical concepts frequently require modification with time, and Dave Holtz's analysis of anionic hyperconjugation is no exception.[177] Physical organic chemistry is a complex science because we generally cannot take a simple "effect" and study it in isolation. We must deal with real compounds in which any change in structure introduces a number of "effects". As a result, interpretations are rarely definitive and noncontroversial. In the present case, a number of theoretical calculations have been interpreted as showing the importance of anionic hyperconjugation. In orbital terms, anionic hyperconjugation is an interaction of a filled lone-pair orbital with a σ* orbital. The corresponding bond should then become longer and weaker. Extensive computational studies of systems of the β-fluoroethyl anion type show exactly such a phenomenon: a C–F bond anti to a carbanionic lone pair is longer than a corresponding bond perpendicular to the lone pair.[178,179] Considerations of electron density distributions suggest a transfer of charge as indicated by Figure 10 although integrations of electron density suggest that any such changes are small.[180–182]

My present view in resolving what appear to be conflicting results is that the charge-transfer resonance structure in Figure 10 is not a satisfactory symbol for the phenomenon. Hyperconjugation between lone pairs and $\sigma^*$ orbitals undoubtedly occurs and results in C–F bonds that are longer and weaker and with the stereochemical effects given by computations. For example, a crystal structure of a $F_3CO^-$ salt shows a short C–O and long C–F bond lengths exactly as expected for such hyperconjugation.[183] However, C–F bonds are already so polar in the sense $C^+ F^-$, that such hyperconjugation results in little additional charge transfer. As a result, the polar effect of CF groups is approximately that given by normal inductive effect considerations and little enhancement of such effects is given by hyperconjugation. Thus, negative hyperconjugation might be expected to give an upfield shift for NMR signals of atoms adjacent to carbanionic sites and not the downfield shifts actually found for $^{19}F$ and $^{13}C$ signals at $CF_3$ in $(CF_3)_3C^-$.[184] A further discussion of hyperconjugation with lone pair electrons is given later (*see* section titled Electron Density Functions).

My awareness of the often dramatic effects of fluorine substitution overlapped with my earlier interest in carbonium ion chemistry and displacement reactions to lead to interesting results in these areas. Research at the 3M Company showed that perfluoroalkanesulfonate esters solvolyze at greatly enhanced rates compared to tosylates.[185] I wondered whether such new leaving groups would rival the nitrogen of diazonium salts and lead to carbonium ion type chemistry from primary alkyl or phenyl esters. Trifluoromethanesulfonic acid or its derivatives were not available commercially at that time, but I received a generous gift of the barium salt from the 3M Company from which we prepared the ethyl ester ("triflate") and deuterium-substituted derivatives. We found that although ethyl triflate solvolyzes more than $10^4$ faster than the tosylate, the isotope effects indicate that to a hydroxylic solvent is significant. This condition undoubtedly holds for the bridgehead carbanions from **10** and **11** because these carbanions are localized and steric hindrance is not a factor. Such acids, in which the conjugate anion has the basic lone pair essentially localized on a single atom and in which there is relatively little structural change between acid and conjugate base have been termed "normal".[172] Other examples of normal organic acids are phenylacetylene and chloroform.[173]

The equilibrium acidities of these bridgehead compounds toward CsCHA in cyclohexylamine were measured by Gene R. Ziegler. Gene had done earlier kinetic acidity measurements during his graduate studies in the middle 1960s. The group at that time had a great deal of group spirit due in no small measure to Gene's ebullient personality. One of the products of that period was our group mascot, "Peter Grivich". Someone in the group had noted a letter in a local newspaper from a 10-year-old boy from the midwest expressing his total disappointment on a visit to San Francisco that the famed "Golden Gate Bridge" was in fact only red. Something about this dashing of youthful expectations touched the emotions of the group, and they adopted the boy's name as our group pseudonym.

Gene Ziegler, of course, was a key figure in all of this. After his Ph.D. he worked for DuPont for several years, then left to teach chemistry at the secondary school level in the Wilmington–Philadelphia area. During one of his summer vacations, he returned to Berkeley and did the equilibrium measurements that combined so well with Dave Holtz's kinetic studies. Gene and a postdoctoral researcher in the group, George Sonnichsen, married sisters. George has stayed with DuPont whereas Gene and his wife, also a schoolteacher, now teach at various schools in exotic locations (e.g., Karachi, Lima, and Sao Paolo) and combine their profession with their love for travel. In talking with Gene it seems clear that his deep knowledge of physical organic chemistry has enhanced his teaching of precollege chemistry. I believe that his Ph.D. and his ability to think have added unique depth to the education courses he has also taken.

One result of this type of kinetic analysis is a method of using isotope effects to determine the amount of internal return in some proton-transfer reactions. In Cram's internal return mechanism, when $k_{-1}$ and $k_2$ are of comparable magnitude, the effect on

$k_{exp}$ in eq 17 is now more complex. The primary isotope effect, for example, has an intermediate value. This effect is seen in the exchange reactions of arylmethanes in methanolic sodium methoxide. For example, $k_D/k_T$ = 1.3 for triphenylmethane at 100 °C.[153] Comparing the rate of introduction of deuterium with the loss of tritium from the reaction of triphenylmethane-*t* in CH$_3$OD gives the ratio $k_H/k_T$, which is equal to 1.7 at 100 °C. We can now use the comparison of both primary isotope effects to derive the degree of internal return. For triphenylmethane-*t*, for example, $k_{-1}/k_2$ = 0.66. This method has important limitations (it does not work, for example, if tunneling is important) but can give useful measures of internal return within given ranges. The results show, for example, that some strongly basic carbanions can react with methanol at rates competitive with solvent diffusion; however, the resulting magnitudes of internal return have negligible effect on the Brønsted correlation discussed already.[153]

When $k_{-1} \gg k_2$, the degree of internal return can only be determined qualitatively as having a large magnitude. In such cases, however, for a series of related compounds for which $k_2$ is expected to have about the same value, the ratio $(k_1/k_{-1})$ is an equilibrium acidity and the experimental relative rates are then measures of this equilibrium acidity. This principle was demonstrated for the fluorinated bicyclics, **10** and **11**. Equilibrium p*K* values were measured for the cesium salts in cyclohexylamine. The compounds are all relatively acidic with p*K* values in the region of 18–23. A Brønsted correlation of the exchange rates compared to the relative ion-pair acidities gives a slope close to unity.[170]

The relative rates and equilibrium acidities can also be completely accounted for on the basis of simple electrostatic considerations. The high polarity of the C–F bond in the sense C$^+$ F$^-$ stabilizes a nearby carbanion without recourse to anionic hyperconjugation. The carbanion from **11** is more stable than the analogous bridgehead carbanion, **12**. This could be interpreted as indicating anionic hyperconjugation because a double bond at the bridgehead in **12** is more strained than that from **11**. But there is an alternative and simpler explanation: in the anion from **11** there are now two additional C–F dipoles that can stabilize the charge by simple electrostatics! Recently, quantitative acidity measurements have been reported for CH(CF$_3$)$_3$ and **10** (X = F) in the gas the solvolyses are still $S_N2$-like in character, even in relatively nonnucleophilic solvents.[186,187] Thus, the triflate group behaves as a "normal" but highly reactive sulfonate leaving group. Since then, of course, the triflate leaving group has become common in organic chemistry.

The chemistry of fluorinated compounds plays a continuing role in my research. For example, pentafluorobenzene was found to be rather acidic; its p$K$ is in the mid-20s on the cyclohexylamine ion pair scale,[188] and both the lithium and cesium salts were shown to form contact ion pairs in cyclohexylamine.[189] This is a surprisingly high acidity for an aryl hydrogen. More recently, the cesium ion pair p$K$ values of various fluorinated and chlorinated benzenes have been determined in tetrahydrofuran (THF).[190] The results can be analyzed in terms of an additive effect per substituent. For the fluorinated benzenes, for example, the effects on p$K$ of an ortho, meta, or para fluorine are –5.2, –3.0, and –1.4, respectively (partial equilibrium factors). These numbers are similar to those for the exchange reactions of fluorinated benzenes with methanolic sodium methoxide[191] and of fluorobenzene with LiCHA.[192] From these results the cesium ion pair p$K$ of benzene can be readily extrapolated as 44.8; the corresponding value for the lithium salt is 39.5.

The whole story of my attending the meeting in Munich, meeting Professor Tatlow, and our ensuing collaborative research shows the value of occasionally participating in meetings that represent potentially new areas for one's research.

## Carbanions in Tetrahydrofuran

These examples show that kinetic and equilibrium studies of carbanions have involved a variety of chemical concepts and types of experimental results. Much of this chemistry has been reviewed.[144,146] Carbanions play an important role in organic synthesis but only rarely in solvents of the type of cyclohexylamine. Since butyllithium became commercially available over two decades ago, one of the most important types of preparative procedures in organic synthesis is metallation of a suitable substrate by an alkyllithium or lithium amide in an ether, often THF, followed by reaction with a suitable electrophile (eq 18).

$$RH \xrightarrow{BuLi} RLi \xrightarrow{E^+} R-E \qquad (18)$$

The electrophile is generally either an alkyl halide or sulfonate (alkylation reaction) or a carbonyl compound (aldol or Michael addition reaction). This chemistry is almost entirely the chemistry of ion pairs and aggregates of lithium derivatives of carbonyl compounds and related functions such as oxime ethers, hydrazones, carboxylates, and corresponding dianions. Many such metallation conditions involve kinetic acidities, but thermodynamic effects are still of significance in such cases. This chemistry moreover involves structure, stereochemistry, and mechanism; it is now a major topic in physical organic chemistry and will undoubtedly continue to be so into the next century. The understanding of this complex chemistry is important for efficient synthesis design. For this reason much of our recent work on carbon acidity has involved organoalkali equilibria and reactions in THF.

In 1965, Hogen-Esch and Smid[193] interpreted the spectra of alkali metal salts of fluorene in THF in terms of an equilibrium of two types of ion pairs, a "tight" or "contact" ion pair (CIP) and a "loose" or "solvent-separated" ion pair (SSIP). This interpretation has been amply confirmed and has been of major importance in understanding ion-pair carbanion chemistry.[194] The CIP is considered to be simply a pair of ions in close juxtaposition, but the SSIP is considered to have a solvent shell around the cation. The small lithium cation prefers to form SSIPs, whereas the large cesium cation forms exclusively CIPs. This contrast is readily explained on a simple electrostatic basis with the help of Figure 12.

The electrostatic energy of two charges varies as the reciprocal of the distance of separation, $1/r$, whereas that of a charge and a dipole varies as $1/r^2$. Thus, several solvent dipoles could compensate for the increased ionic separation in an SSIP to make it the more stable structure. In THF the two structures are frequently comparable in energy for the small lithium cation, particularly for highly delocalized carbanions where the effective

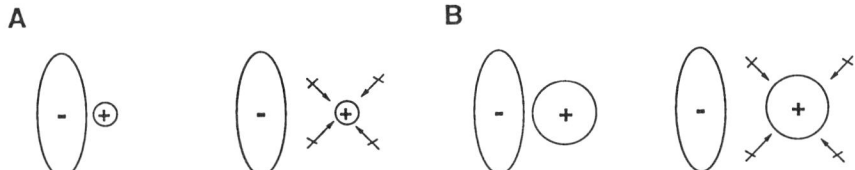

*Figure 12. Diagrams of contact ion pairs (CIPs; left) and solvent-separated ion pairs (SSIPs; right) with small and large cations. A: delocalized anion with small cation; B: delocalized anion with large cation.*

charge–charge distance is already large by comparison with the lithium cation–solvent dipole distance. For a large cation, however, the dipole stabilization falls off so rapidly with distance that it can no longer compensate for the reduced charge–charge interaction in the SSIP.

The different ion pairs are often characterized by changes in their electronic spectra. In a carbanion, the excited state usually involves transfer of electron density from the center to the periphery. In an ion pair with the cation located close to the center of charge in the anion, such charge transfer away from the cation is more difficult, and absorbance occurs at higher frequency than in the free ion. In a CIP, the lithium cation is closer to the anionic charge than a cesium cation, and lithium CIPs absorb at shorter wavelength than cesium CIPs. For lithium SSIP salts, however, the reverse is true. Sodium salts are intermediate in behavior; they show some tendency toward SSIP, whereas the larger potassium salts are generally CIPs.

Our first step was to set up indicator acidity scales for cesium and lithium salts in THF.[195–197] The early work was done by three top graduate students (Michael J. Kaufman, Daniel A. Bors, and Scott Gronert) and has been refined in recent work. The resulting scales are compared with other results in Table III. With a few exceptions the numbers all agree rather well. Most of the p$K$ values shown for lithium compounds pertain to the solvent-separated ion pairs. It seems clear that in general the changes in structure have about the same effect on ion-pair interactions with the large cesium cation in a CIP, a lithium cation in its SSIP, and the free ion in DMSO. In Table III it is important to remember that only the DMSO values are absolute; they refer to the dilute DMSO solution as the standard state. The ion-pair values are relative and

Table III. Comparison of Acidities in Different Media Normalized to the DMSO Value for Fluorene

| Compound | DMSO[a] | LiCHA[b] | CsCHA[c] | THF[d] Li+ | THF[e] Cs+ | DME[f] Li+ | DME[g] Cs+ | THF[h] Li+(Crypt) |
|---|---|---|---|---|---|---|---|---|
| 1,3-Diphenylindene | | 13.7 | 13.5 | 12.3 | | | | |
| 1,4-$C_5H_4Ph_2$ | 14.6 | | 14.2 | | | | | |
| Cyclopentadiene | 17.6 | | 16.1 | | | | | 21.3 |
| 7-Ph-benz[c]fluorene | | | 15.5 | 14.9 | 15.6 | | | 16.1 |
| 9-Phenylfluorene | 17.9 | 18.5 | 18.35 | 17.6 | 18.15 | 17.75 | 18.15 | 18.5 |
| 2-Phenylindene | 19.4 | | | | | | | 20.2 |
| Benz[a]fluorene | | 20.3 | 20.2 | 19.7 | 20.1 | | | |
| Indene | 20.1 | 20.3 | 19.8 | | | 19.55 | | 21.7 |
| Benz[c]fluorene | | 19.7 | 19.6 | 19.3 | 19.5 | | | |
| Benzanthrene | | 21.0 | 21.3 | 20.1 | | 20.15 | 21.25 | |
| 9-Benzylfluorene | | | | | 21.3 | | | 22.0 |
| 9-Methylfluorene | 22.3 | 22.65 | 22.2 | 22.5 | 22.3 | | | |
| Fluorene | 22.9 | [22.9] | [22.9] | [22.9] | [22.9] | [22.9] | [22.9] | [22.9] |
| Benzo[def]fluorene | 22.5 | 22.9 | 22.8 | | 22.9 | | | |
| PhCH$_2$CN | 22.2 | 17.7 | 22.7 | | | | | 23.2 |
| Benz[b]fluorene | 23.5 | 23.5 | 23.3 | 23.0 | 23.6 | | | |
| 9-tert-Butylfluorene | 24.35 | 24.9 | 24.1 | 24.4 | 24.4 | 24.1 | 24.35 | |
| 2-(Dimethylamino)-fluorene | 24.5 | | | | | | | 24.5 |
| Tetraphenylpropene | 25.8 | 25.95 | | | | 24.8 | | |
| Triphenylpropene | 25.9 | 26.45 | 26.45 | | 26.8 | | 26.15 | 26.4 |
| Ph$_2$CHSPh | 26.8 | | | | | | | 27.5 |

# A Lifetime of Synergy with Theory and Experiment

| Compound | | | | | | |
|---|---|---|---|---|---|---|
| 9-Phenylxanthene | 27.9 | | 28.35 | | 28.7 | 27.0 | 28.3 | 28.3 |
| Phenylacetylene | 28.7 | 23.2 | | | | | | 31.7 |
| p-Biphenylylacetylene | | | | 22.0[i] | | | | |
| Triphenylpropyne | | | | 22.4 | 29.6[i] | | | |
| p-Biphenyl-diphenylmethane | 29.4 | | 30.0 | | 30.1 | 28.4 | 29.9 | 29.9 |
| Xanthene | 30.3 | | | | | | | 30.3 |
| Triphenylmethane | 30.6 | | 31.3 | 31.0 | 31.4 | 29.95 | 31.35 | 31.4 |
| Tri-p-tolylmethane | | | 32.9 | | 33.1 | | | |
| Diphenylmethane | 32.5 | | 33.2 | | 33.25 | | | 33.0 |
| Di-o-tolylmethane | | | 34.7 | | 34.2 | | | |
| Di-m-xylylmethane | | | 36.2 | | 36.0 | | | |
| p-Methylbiphenyl | | | 38.9 | | 38.7 | | | |

NOTE: p$K$ values are given per hydrogen.

[a] DMSO is dimethyl sulfoxide.
[b] Lithium cyclohexylamide (LiCHA) in cyclohexylamine.
[c] Cesium cyclohexylamide (CsCHA) in cyclohexylamine.
[d] Tetrahydrofuran (THF) with lithium gegenion.
[e] THF with cesium gegenion.
[f] Dimethoxyethane (DME) with lithium gegenion.
[g] DME with cesium gegenion.
[h] THF with lithium cation as the [2.1.1]-cryptate.
[i] At $10^{-4}$ M; salt is aggregated.

SOURCE: Data are from the following references: [a]198; [b]144 and 146; [c]144 and 146; [d]196 and unpublished results; [e]197; [f]199; [g]199; [h]199–203; [i]204.

are given an absolute basis only by choosing one compound, fluorene, as a standard and assigning the absolute DMSO value to it. The numbers in Table III are also corrected for statistics and give the acidities per hydrogen; for example, the actual pK of fluorene is lower by log 2 (0.3) because it has two equivalent acidic hydrogens, whereas that for indene is unchanged because it has two acidic hydrogens and two equivalent carbons to which the proton can return.

A few numbers show large differences. The LiCHA pK for phenylacetylene was cited earlier to be substantially lower than that in DMSO. The pK could not be determined for the cesium salt for solubility reasons. In THF a comparison of Li and Cs pK values was also difficult either because of aggregation or limited solubility. Recently, however, we were able to show that both lithium and cesium salts of 1-ethynyladamantane are soluble in THF and monomeric in dilute solution; the pK values are 23.7 and 31.6, respectively.[204] Thus, the lithium salt is undoubtedly a CIP; the pK difference is a quantitative measure of the enhanced electrostatic interaction of the smaller cation to the localized carbanion. This comparison can be extended in the other direction. The LiCHA pK for phenylacetylene is much lower than the value for a lithium cation encrypted as the [2.1.1]-cryptate. The latter is equivalent to a SSIP. The difference between the DMSO value and the cryptated lithium pK may well indicate that the acetylide ion is appreciably ion paired even in DMSO. Cyclopentadiene also shows significant differences. The lower value with CsCHA may indicate stronger ion-pair association for this less delocalized carbanion. The Russian group[199-203] has argued that the substantially lower value in DMSO compared to the lithium cryptate also results from ion pairing in DMSO, but I suspect that the cryptate result is more likely in error perhaps because of dimerization of cyclopentadiene during the measurement.

Our acidity studies in THF have thus resulted in useful quantitative scales of relative ion-pair acidities for a growing variety of compounds. Conductivity measurements of the lithium and cesium salts of the indicator systems in THF have also given important results.[205] The dissociation of the lithium salts is almost 3 orders of magnitude greater than that of the cesium salts as expected for a comparison of SSIPs with their reduced ionic interactions with CIPs. The resulting dissociation constants to the free ions also provide the corresponding relative "free ion" acidities in this solvent. These ionic acidities are, of course, the same for lithium and cesium and also compare well with the relative acidities in DMSO.[205] This direct comparison of the lithium and cesium scales shows that there are no significant systematic errors and that these scales are now suitable for study of other types of compounds. A number of these applications have already been made.

An example is shown with the compounds in Table IV. Sulfur is now generally considered to stabilize an adjacent carbanion by a polarization mechanism rather than by delocalization of charge via its empty d orbitals (*see* section titled Electron Density Functions). Selenium is less effective in such stabilization despite its greater polarizability probably because the C–Se bond is longer than the C–S bond. The electrostatic energy of a charge and a polarizable center falls off as the fourth power of the distance. A phenylthio group is almost 3 pK units more effective than a methylthio group undoubtedly because of the greater inductive effect and polarizability of the aromatic ring. Further extension of the ring to the biphenylylthio group has but a small additional effect. For the directly conjugated phenyl ring, however, changing to a *p*-biphenylyl group causes a significant lowering of the pK by 1.6 units.

Table IV. Cesium Ion-Pair Acidities of Some Benzyl Sulfides and Selenide in THF

| Compound | pK |
|---|---|
| Ph–CH$_2$SPh | 31.8 |
| Ph–CH$_2$SePh | 33.2 |
| Ph–CH$_2$SCH$_3$ | 34.7 |
| Ph–Ph–CH$_2$SPh | 30.2 |
| Ph–CH$_2$S–Ph–Ph | 31.5 |

NOTE: pK values are per hydrogen.
SOURCE: pK values are from reference 206.

Table V. Cesium Ion-Pair Acidities of Some Dithianes in THF

| Compound | pK |
|---|---|
| 1,3-dithiane | 36.5 |
| 2-methyl-1,3-dithiane | 38.2 |
| 2-phenyl-1,3-dithiane | 30.5 |
| 2-(p-biphenylyl)-1,3-dithiane | 29.1 |
| 2-methyl-2-(1,3-dithian-2-yl) bicyclic trithio | 30.2 |

NOTE: pK values are per hydrogen.
SOURCE: pK values are from reference 207.

Extension of these results to comparable dithianes is shown in Table V. A single RS group was shown in Table IV to reduce the pK of toluene (41.2)[197] by 6.5 pK units. A second RS group as in phenyldithiane gives a further reduction of 4.2 units. Phenyldithiane is more acidic than triphenylmethane (pK = 31.3), and changing to a *p*-biphenylyl group gives a further reduction of pK of 1.4 units. The additional sulfur in the bicyclic compound shown in Table V gives an effect comparable to the phenyl substituent. In fact, the stabilization of a carbanion by sulfur is not much different from that of phenyl. The ion-pair pK of dithiane itself (36.5) is not much higher than that of diphenylmethane (33.25).[197] Similar comparisons have been found by the Bordwell group in DMSO.[198]

Table III contains few pK values of lithium salts in THF above the middle twenties because proton transfers in this region are inconveniently slow. By using a new technique involving multiple equilibria that include cesium salts, we have recently been able to obtain lithium pK values for 2-phenyl-1,3-dithiane (29.4 for the lithium CIP) and 2-*p*-biphenylyl-1,3-dithiane (28.2, CIP; 29.3, SSIP).[208] The latter SSIP value is close to the cesium pK listed in Table V, and the lithium CIP values are somewhat lower.

The dithiane carbanions without a delocalizing aryl substituent do not absorb in the useful UV region and had to be determined by the so-called "single-indicator" technique. In this method, the absorbance of an indicator is measured and a known amount of substrate is added. The resulting decrease in the indicator absorbance then tells how much of the substrate was converted to its anion. This technique makes great demands on purity and technique and has recently been greatly improved.[208] It has led in the past to some systematic errors. Our recent work has been aided greatly by a unique facility, a glovebox–spectrometer combination. In our most recent version, a sample well built into the floor of a glovebox contains a thermostated cell holder with quartz fiber-optic cables connected to a dedicated external UV–visible spectrophotometer. Thermostating is controlled by an external circulating bath. Materials are stored in the glovebox after purification, and all manipulations and measurements are made in the glovebox. Because the time interval between solution preparation and spectral measurement is short, even solutions of relatively unstable carbanions can be examined. Much of the work accomplished recently would not have been possible without such a facility.

Some results of sulfones and sulfoxide are also of interest (Table VI).[209] The cesium ion pair p$K$ of benzyl phenyl sulfone, 23.0, is similar to the ionic p$K$ in DMSO, 23.7.[198] Changing the conjugating phenyl group to $p$-biphenylyl has a much smaller effect, 0.7 p$K$ units, than in the sulfides and dithianes discussed previously. The explanation is probably in the strong electron-attracting inductive effect of the sulfone group that inhibits delocalization of charge to remote regions. The sulfoxide group is clearly less stabilizing than the sulfone substituent. These compounds were also studied with lithium as the gegenion and, as shown in Table VI, the lithium p$K$ values are 4–5 units lower

Table VI. Cesium and Lithium Ion-Pair Acidities (Per Hydrogen) of Some Sulfones and a Sulfoxide

| RH | p$K$ (Cs) | p$K$ (Li) |
|---|---|---|
| PhCH$_2$SO$_2$Ph[a] | 23.0 | 19.5 |
| $p$-PhC$_6$H$_4$CH$_2$SO$_2$Ph | 22.3 | 18.8 |
| $p$-PhC$_6$H$_4$CH$_2$SOPh | 25.2 | 20.1 |

[a]DMSO value is 23.7 (from reference 198).
SOURCE: Data are from reference 209.

than cesium. As mentioned already, the lithium salts of the indicators used exist primarily as SSIPs, but the lithium salts of these sulfur compounds are CIPs probably by strong association with the oxygens on the sulfur. The stronger ion-pair associations were also demonstrated by some conductivity studies.[209]

In a recent result we found the cesium ion-pair pK of benzylferrocene to be 35.7.[210] This value is only somewhat higher than diphenylmethane (33.25) and indicates that the ferrocene ring stabilizes an adjacent carbanion but is somewhat less effective than a phenyl group.

Water is always a concern. It became a particular concern when Todd McDermitt, a student working with Clayton Heathcock on synthetic applications of aldol addition reactions, used our facilities to do some measurements of the kinetics of the reaction. Todd is an excellent experimentalist but kept getting nonreproducible results. The problem was finally traced to the effect of minute amounts of water in the tetrahydrofuran, and his work led to some important findings. Our normal purification method results in THF containing about $10^{-3}$ M water. Storing the solvent in the glovebox over molecular sieves brings the water down only to $10^{-4}$ M. One limitation appears to be the use of Pyrex glassware in which water is extracted slowly from the glass even after prolonged baking. We have discovered that with the use of quartz vessels the water content can be reduced to a level so low we cannot measure it, $<10^{-6}$ M.

One of the important outcomes of these studies in THF has been a method for determining aggregation numbers in dilute solution, down to $10^{-5}$ to $10^{-3}$ M. Other methods have established that some lithium enolates are aggregated in more concentrated ethereal solutions, of the order of 0.1 M and higher.[211-215] Our new method developed from the discovery that the experimental pK of acetophenone as measured with the cesium salts of indicator hydrocarbons varies systematically with the concentration of the cesium enolate. This variation is a direct result of aggregation of the cesium enolate; in effect, the aggregation equilibrium (eq 20) pulls the acid–base equilibrium of eq 19 to the right.

$$\text{In}^- \text{Cs}^+ + \text{PhCOCH}_3 \rightleftharpoons \text{InH} + \text{Ph}\!-\!\!\!\!<\!\!\!\begin{array}{c}\text{O}^-\text{Cs}^+\\\phantom{x}\end{array} \quad (19)$$

$$n \, \text{PhCOCH}_2^- \, \text{Cs}^+ \rightleftharpoons (\text{PhCOCH}_2^- \, \text{Cs}^+)_n \quad (20)$$

The slope of a plot of the observed pK vs. the logarithm of the cesium enolate concentration is equal to $(1 - \bar{n})/\bar{n}$, where $\bar{n}$ is the average aggregation number.[216] For acetophenone such a plot over a concentration range of $2 \times 10^{-5}$ to $5 \times 10^{-3}$ M is a straight line with $\bar{n} = 3.5$. Studies of this type have indicated that the more delocalized the charge or, correspondingly, the less basic the enolate ion, the less the degree of aggregation.[217,218] For example, dibenzyl ketone gives a curved plot (Figure 13) in which the slope ranges from a value of almost 2 at high concentrations to close to unity at low concentrations. From the results the equilibrium constant for dimerization can be evaluated, $K_{dimer} = 1890 \, \text{M}^{-1}$.[218]

For the simple case of a monomer–dimer equilibrium, the data can be plotted in a different way. This method gives a straight line from which the constants can be obtained simply and precisely (Figure 14).

This work was done by a postdoctoral co-worker, Jim Ciula, who also made the important observation that the $\lambda_{max}$ of several cesium enolates change significantly with concentration; there is a shift of several nanometers to shorter wavelengths at higher concentrations.[118] These observations were followed up by Jim Krom, one of my last graduate students, a careful and astute

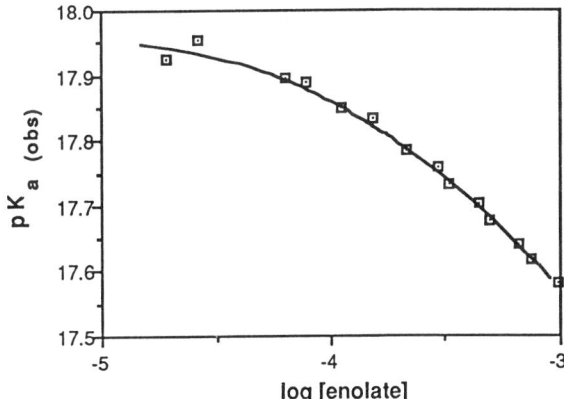

*Figure 13. Aggregation plot for the cesium enolate of 1,3-diphenylacetone. The curve shown is the theoretical curve for $pK_a$ of the monomer = 17.97 and $K_{dimer}$ = 1890 $M^{-1}$. (See reference 218.)*

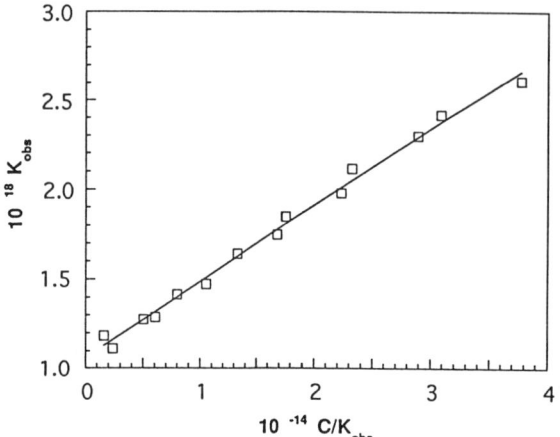

*Figure 14. Linear aggregation plot for the cesium enolate of 1,3-diphenylacetone. The slope is $2K_{dimer}K_a^2$ and the intercept is $K_a$. The line shown gives $pK_a$ of the monomer = 17.97 and $K_{dimer}$ = 1885 M$^{-1}$. C is the concentration of enolate; $K_{obs}$ is experimental equilibrium ion pair acidity. (See reference 218.)*

chemist with a good mathematical sense. He showed that different aggregation states of some lithium and cesium salts have measurably different UV–visible spectra; the changes in equilibrium distribution of these aggregates with concentration then give significantly different net spectra with changes in concentration. Jim demonstrated how the technique of "singular value decomposition" (SVD) can be applied to a series of such spectra.

For a system of two components, for example, the SVD procedure gives a series of vectors in the multidimensional space of the absorption wavelengths. The first vector is the best representation of both components in a least-squares sense, the second gives the first-order deviation, and the remaining vectors describe noise. From the analysis one can recover the spectra of the two individual components. By comparison of this method with the previously discussed method, the change in apparent pK with concentration, one can determine not only the degree of aggregation but also the number of monomer units in the aggregates. This method was applied recently to lithium and cesium diphenylamide in THF.[219] The cesium salt was found to be slightly aggregated with the aggregates being dimers and $K_{dimer}$ = 160 M$^{-1}$. The pK of the monomer, 24.2, compares well with the free ion pK in DMSO, 24.95,[198] in agreement with the general correspondence of

## A Lifetime of Synergy with Theory and Experiment

the cesium ion pair p$K$ values in THF and DMSO. The lithium salt was found to be monomeric with a p$K$ of 19.05; conductivity studies as well as the spectra show lithium diphenylamide to be a CIP in THF. The substantial negative charge on nitrogen clearly favors CIP formation with the small lithium cation.

We are currently applying these methods to study the actual aggregation state in alkylation and aldol addition reactions. The complexity of the general system can be gauged by the summary in Scheme IX; that is, the simple conversions given in eq 18 are oversimplified. Each of the components in the complex equilibria of alkali–organic compounds [free ions, monomeric ion pairs (CIP and SSIP), and aggregates of different types] will have its own rate constant and stereochemistry in reaction with a given electrophile. Determining the actual reactant in such a series of rapid equilibria is a difficult problem requiring careful equilibrium and kinetic studies. It is also a general problem in chemistry. Examples include the reactions of conformational isomers identified with the names of Curtin, Hammett, Winstein and Holness, and reviewed by our editor.[220a]

For the present case of reactions of lithium enolates, a mechanism has been proposed[220b] in which the reactant is the known cubic tetramer. Unfortunately, this mechanism has virtually no evidence in its support. The actual reactant might well be a small amount of a lower aggregate in equilibrium with the dominant tetramer. We have started working on this problem and have already obtained some significant results. Indeed, the cases that we have studied so far show that alkylation reaction occurs dominantly via the small amount of monomeric ion pairs in equilibrium with much larger amounts of higher aggregates.

In a reaction of a metal enolate that is an equilibrium mixture of various aggregates, the logarithm of the experimental

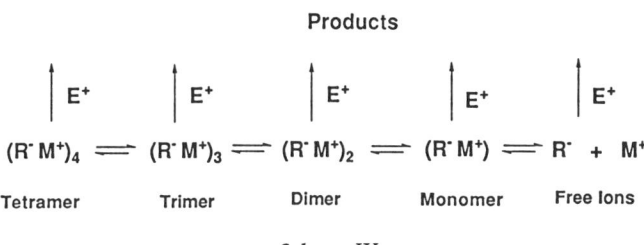

Scheme IX.

initial rate plotted against the logarithm of the formal enolate concentration gives a curve whose slope at any point is $\bar{n}^{\ddagger}/\bar{n}$, where $\bar{n}$ is the mean degree of aggregation of the metal enolate and $\bar{n}^{\ddagger}$ is the mean number of metal enolate units in the reaction transition state. Since the quantity $\bar{n}$ can be determined by the equilibrium measurements mentioned earlier, the kinetic studies then give the desired quantity $\bar{n}^{\ddagger}$.[221a]

This method was applied to alkylations of the cesium salt of *p*-phenylisobutyrophenone, **13** (eq 21). Reaction with methyl tosylate, for example, gives almost equal amounts of **14** and **15** (R = CH$_3$).

$$\mathbf{13} \xrightarrow{RX} CsX + \mathbf{14} + \mathbf{15} \quad (21)$$

The *p*-phenyl substituent in **13** provides a sufficiently strong chromophore to permit direct spectroscopic measurement. A plot of the experimental pK vs. the logarithm of the formal enolate concentration gives Figure 15, in which the slope of the line shown, −0.690, corresponds to a mean degree of aggregation of 3.22. The spectrum changes significantly with concentration. More recently, analysis of the spectra over a wide concentration range has shown the presence of monomers, dimers, and tetramers and has allowed determination of the equilibrium constants between them.[221b]

Kinetic measurements were made for the reaction of **13** with methyl tosylate. Initial rates were used to avoid possible complications of mixed aggregates with the cesium tosylate formed. From these results $\bar{n}^{\ddagger}$ is close to unity; that is, although the enolate is present mostly as dimers and tetramers, the alkylation involves predominantly the monomeric ion pairs. This result differs somewhat from our preliminary report.[221a] It was only after

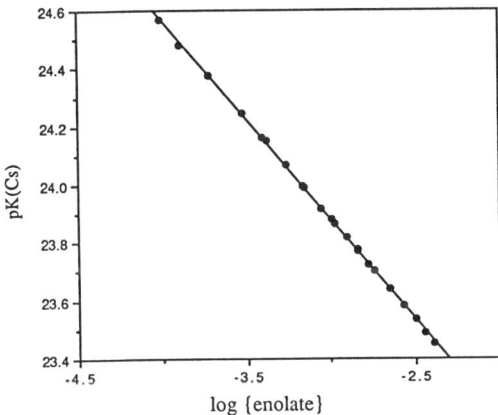

*Figure 15. The experimental cesium ion pair pK of **13** at 25 °C plotted against the logarithm of the formal enolate concentration. The line shown has a slope of –0.690.*

that report that we discovered that the spectra of this enolate is one of those that vary significantly with concentration, and this factor needs to be taken into account in analyzing aggregation and the kinetics.

Study of the corresponding lithium enolate is more difficult because its spectrum overlaps with that of the unreacted ketone. Its $\lambda_{max}$ is shifted to much shorter wavelengths from the effect of the nearby lithium cation. Recently, we have been able to determine the spectra of the lithium salt of $p$-phenylisobutyrophenone in the absence of interfering ketone. These spectra showed a change in $\lambda_{max}$ from 336 nm at $3 \times 10^{-3}$ M to 342 nm at $3 \times 10^{-4}$ M. Application of the SVD analysis showed the presence of two components in the ratio monomer–tetramer having $\lambda_{max} = 343$ and 331 nm, respectively. Interestingly, the amount of dimer present was too small to be observed. The pK of the lithium salt as determined using the "single indicator technique" is a curved function of log[enolate] and also shows that this salt is mostly monomer at $10^{-4}$ M and tetramer at high concentrations (Figure 16).[223] The curve shown in this figure is that derived from the spectral measurements with $K_{tetramer} = [tetramer]/[monomer]^4 = 5.2 \times 10^7$ M$^{-3}$. These results show that the monomer content at typical synthesis concentrations is about 1–2%, high enough to be of kinetic importance.

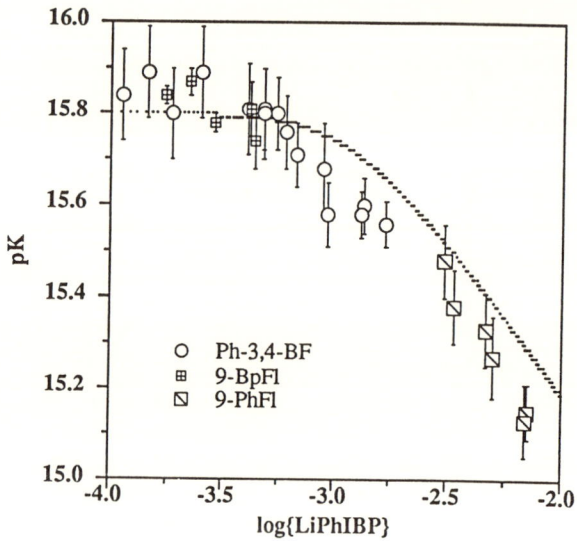

*Figure 16. Aggregation plot for LiPhIBP, the lithium salt of p-phenylisobutyrophenone, in THF at 25 °C. The curve shown is the theoretical curve for $pK_a$ (monomer) = 15.80 and $K_{tetramer} = 5.2 \times 10^7\ M^{-3}$. Three indicators were used: 9-phenyl-3,4-benzofluorene (Ph-3,4-BF), 9-p-biphenylylfluorene (9-BpFl), and 9-phenylfluorene (9-PhFl).*

Another recent study made use of the lithium enolate of p-phenylsulfonylisobutyrophenone, **16**. The sulfonyl group lowers the energy of the excited state sufficiently that $\lambda_{max}$ is now accessible. This compound exists mostly as a dimer in THF (the equilibrium constant $K_d = 5 \times 10^4\ M^{-1}$), but its reaction with methyl tosylate or *p-tert*-butylbenzyl bromide again involves dominantly the monomeric ion pair.[222] Further kinetic analysis shows that the monomeric ion pair is 3000 times more reactive toward the benzyl bromide than is the dimer. Thus, even under synthesis conditions, reaction occurs mostly via the monomer. This work is now being extended to the mixed aggregates with LiBr. Preliminary kinetic analysis shows that the monomer is again much more reactive than its mixed aggregate with LiBr.[222b]

**16**

# A Lifetime of Synergy with Theory and Experiment

These examples suggest that the proposed reaction mechanism of lithium enolate cubic tetramers[220b] may rarely be applicable. The reactions of metal enolates are extremely important in synthetic organic chemistry, and working out the quantitative aspects and general principles of Scheme IX is a critical current problem in physical organic chemistry. I think our recent results are an important addition to this chemistry and, accordingly, I expect that this new work will remain a significant part of my chemical research in the foreseeable future.

One of the other interesting recent results of the past decade was the finding that ion-pair acidities involving dianions can be amazingly high compared to those of the corresponding monoanions. For example, the cesium ion-pair pK of **17** in THF is 2.0 units higher than the pK of the monocesium salt going to the dicesium salt![224] For the lithium salts the $\Delta$pK is only 0.4 units, and the difference can be accounted for in terms of different types of ion pairs. All of the cesium salts are undoubtedly CIPs. The mono-lithium salt is SSIP, but, because of the higher charge of the dianion, in the dilithium salt at least one of the lithiums is a CIP. For 2,2′-biindenyl, **18**, the differences are not as dramatic. The monocesium salt is less acidic than the hydrocarbon, but $\Delta$pK is only 1.1 units.[224] For the lithium compound $\Delta$pK is 2.2 units in THF. In these cases one can hardly generate the monoanion without making some of the dianion. Yet, in other cases the second pK is much higher than the first. For the indenofluorenes **19** and **20** $pK_2$ for the cesium salts in THF are more than 5 units higher than $pK_1$.

Speakers at a Conference on Carbanion Chemistry, Ottawa, Ontario, Canada, July, 1989. From left: Laren Tolbert, Paul Schleyer, Dave Collum, Marye Anne Fox, Charles DePuy, Andrew Streitwieser, Ted Cohen.

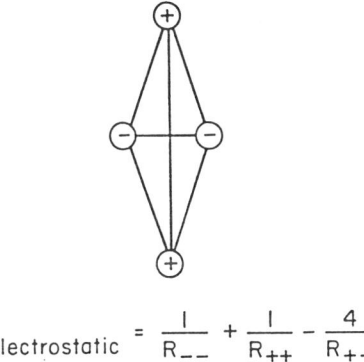

$$E_{electrostatic} = \frac{1}{R_{--}} + \frac{1}{R_{++}} - \frac{4}{R_{+-}}$$

*Figure 17. Coulomb interactions for a point charge model of a dication salt of a dicarbanion. E is the electrostatic energy, and R is the distance between cations or anions.*

Similar results had been reported earlier in cyclohexylamine and were interpreted on the basis of a triple-ion structure with one cation above the ring plane and the other below.[225-227] In a pattern of charges as in Figure 17, the net electrostatic energy is negative for chemically significant bond distances; that is, because of the presence of two cations, the total attraction between cations and anions exceeds the repulsion between like charges even if the carbanion charges are on adjacent atoms. For this type of case the relative ion-pair acidities are much different from those involving the free ions. In DMSO, for example, $\Delta pK$ ($pK_2 - pK_1$) for **19** and **20** is more than 10 units. The difference is small for **17** and **18**, but the dianions in these cases even in DMSO are probably ion paired to one cation.

The proposed triple-ion structure was recently confirmed by NMR studies.[228] The interpretation in terms of elementary electrostatics also has wide ramifications in organic chemistry.[227] A typical example is the demonstration of the remarkably low lithium ion pair acidity of lithium 1-naphthylacetate in THF.[229] An important aspect of this work to me is its demonstration that, despite my undoubted involvement with rather sophisticated theory in organic chemistry, many important experimental results can be interpreted in a completely satisfactory manner with nothing more complicated than Coulomb's law!

# Theoretical Chemistry

## Computers and Molecular Orbital Theory

I realized early in my chemical career that my strengths are not in natural products synthesis. I was more interested in the relationship between structure and reactivity and gravitated naturally toward physical organic chemistry. The study of reaction mechanisms was just one tool in understanding structure and reactivity. Theory in the early years meant resonance structures and their qualitative interpretation. I mentioned in my early history how, even in high school, I was fascinated by the use of resonance structures in interpreting reactions. But it was all so qualitative! Learning to do Hückel molecular orbital (HMO) calculations as a postdoctoral fellow with Jack Roberts was an exciting jump. I had always enjoyed mathematics, but more in applications than in pure theory, and here was a great application. It was exciting fun to see how well the calculated energies compared with experimental chemistry. This enjoyment followed me to Berkeley, and theory has been an important part of my research ever since. The early part started of course with HMO theory.

HMO theory has been exceedingly useful in organic chemistry, particularly in qualitative concepts, undoubtedly because the theory gives the correct nodal behavior of wave functions; the character of nodes, especially for the highest occupied molecular orbitals (HOMOs) and lowest unoccupied molecular orbitals (LUMOs), is sufficient for many interpretations and predictions. Nevertheless, comparison with numerous experimental results, from our laboratories and others, has shown the limited *quantitative* applicability of the Hückel method. Accordingly, I followed and tried out each new development of computational quantum chemistry for its potential quantitative applications to our chemistry. Each step, from self-consistent field (SCF)-$\pi$ calculations to semiempirical methods of the Complete Neglect of Differential Overlap (CNDO) type and beyond, was forced by the limitations of each successive approach. And each step was taken somewhat reluctantly; after all, I was trained as an experimental organic chemist and my knowledge of theory and mathematics cannot compare to that of trained theoreticians. But I did learn the different methods and developed a good physical sense of their relationship to real chemistry. Each of these escalating steps in theory was also accompanied by the developing power of computers.

This power was first suggested to me by the first graduate student who started research with me, Robert H. Jagow, in 1953. About 1955, on his own initiative, he used the IBM-650 computer located at the Lawrence Radiation Laboratory (now Lawrence Berkeley Laboratory) to do some large HMO calculations. The 650 was a magnetic drum computer with limited memory; its total capability was substantially less than my first personal computer of the late 1970s! But it was far more powerful than a mechanical calculator. At that time I knew nothing about computers, and I do not know where Bob got the program for doing his calculations.

Shortly thereafter the campus installed an IBM 701 computer. My group was one of the first in the Chemistry Department to make extensive use of it. The 701 and its successor, the 704, were vacuum-tube computers with magnetic core memories. They were large and required a big room with lots of air conditioning to remove the heat produced by the vacuum tubes. I took a course in "Symbolic Assembly Programming" and learned to write programs for these machines. FORTRAN did not exist at that time, and we learned what is now called machine language. We wrote our own program for doing HMO-type calculations. The program and all of the input data were contained on punched cards. I did a good deal of programming at that time, and with my first postdoctoral student, Dr. P. Madhavan Nair, published our first paper that made use of IBM 701 computations in 1959.[230]

# A Lifetime of Synergy with Theory and Experiment

*John Brauman playing the banjo at a group party in 1961.*

We found simple models for the hyperconjugation of alkyl substituents within the Hückel framework and got excellent correlations with experimental ionization potentials. FORTRAN appeared shortly thereafter, and when John Brauman° joined my research group in 1959 he learned FORTRAN programming from IBM manuals. We redesigned the HMO program, which he rewrote for the IBM 704. This program had some clever features; I used some of these features in the HMO program I wrote in BASIC for IBM-type personal computers, but commercially available programs are now easier to use. Brauman's results were printed in easily readable form, and subsequently we published a large number of HMO computations in book form.[231]

---

° *See* Appendix A for brief biographical information about each person designated with the ° symbol.

*Professor Charles Coulson is one of the pioneers in the development of HMO theory and its applications to organic chemistry among many other contributions to theoretical chemistry. I spent one month at Oxford with him during my Sabbatical leave in 1969.*

During production I learned that Professor Charles Coulson° was preparing a similar but smaller volume, and we collaborated on this publication.[232] Several years later, in 1969, I spent one month at Oxford University with Professor Coulson during a Sabbatical leave. He was a kind man and delighted in showing me around Wadham College and relating some of its history. He died of cancer only a few years later. At the time of my Sabbatical, he was the Rouse Ball Professor of Mathematics and head of the Mathematics Institute that occupied a new building only a few years earlier. I spent some time at the Institute and talked with several of the mathematicians there but found we had little in common. The tables that I published with Coulson were useful for a number of years but are not as useful now because the calculations can be reproduced so readily with personal computers.

One of the first approaches we tried beyond the simple Hückel method was the ω-technique introduced by Wheland and Mann.[233] In this method the effective electronegativity of a π-carbon is made proportional to its π-charge. I showed that the method was an approximation to an SCF-$\pi$ calculation in which electron repulsion is treated explicitly and that it gives a successful account of ionization potentials.[234] We were able to apply it also to solvolysis[86] and aromatic substitution[235] reactions, but the method now is just of historical interest.

Semiempirical self-consistent field $\pi$-methods (SCF-$\pi$) became available in the 1950s through the work of Pariser and Parr[236,237] and of Pople.[238,239] They introduced several important approximations for the electron-repulsion integrals, the neglect of some overlap integrals, and the use of experimental energies for certain other integrals. We were able to apply these methods to the spectra of our indicator hydrocarbon salts[240], to aromatic substitution[241], and to reactions giving arylmethyl cations[86] and anions.[242] These methods were reasonably successful, but they are still limited to $\pi$-electrons only. The first all-valence-electron semiempirical methods, such as CNDO,[243,244] also came out at this time, and we were of course anxious to compare such computations with experimental results. For benzylic cations some quantitative correlations were obtained, but the method was found to have serious limitations.[245] Subsequent study of carbanions showed that the CNDO/2 method gives a poor account, primarily because of its inadequacy in handling lone pairs.[242] The method was particularly poor with alkyl anions.[246] Further analysis showed that a major cause is the necessity within the CNDO method for handling all orbitals as spherically symmetrical for the computation of electron-repulsion integrals. This limitation is required to make the calculations independent of the choice of coordinate axes. The result is an exaggeration of repulsion and a spreading of the anionic charge to the peripheries of the carbanions. These disappointing results of the semiempirical methods then available made me receptive to ab initio quantum chemistry.

## Ab Initio Quantum Organic Chemistry

In the winter quarter of 1968, I attended a special seminar course held jointly with the San Jose Research Laboratory of IBM on "Quantum Theory and Computations of Molecular Electronic Structure". This course taught me that ab initio MO calculations are not beyond the capabilities of organic chemists. The differences are significant. Semiempirical methods treat only the valence electrons and use semiempirical values for most of the integrals required; many of these integrals are set to zero. In ab initio (L., from the beginning) methods all electrons are considered and all of the many integrals required are computed. Large computers are therefore required for molecular computations.

Through the cooperation of Dr. Enrico Clementi and his associates at IBM, we obtained a copy of their Hartree–Fock MO program, IBMOL4. By then the campus had a CDC (Control Data Company) 6400 computer, but, of course, IBMOL4 was designed to run on IBM computers. Fortunately, Peter Owens started in my research group at that time. Pete had transferred from Oregon State University to do experimental physical organic chemistry at Berkeley, and he did do extensive work in kinetic studies of carbon acidity. But he also discovered the opportunity to do computational quantum chemistry at Berkeley and his own enjoyment and abilities in this area. He was largely responsible for modifying and improving IBMOL4 for the CDC computer, and by mid-1969 we were doing a number of computational experiments and learning about basis sets. We were joined in these studies by a new postdoctoral student, Richard A. Wolf, and were helped greatly by generous donations of computer time from the campus Computer Center and by John Pople°, who sent us his STO-NG basis sets before publication. This work led to our first ab initio paper in 1970.[247] Even at that time it was clear to me that one of the important types of contribution of ab initio computations to physical organic chemistry lies in its ability to give meaningful results with idealized reference systems for testing chemical concepts such as inductive effects, hyperconjugation, and field effects.[248]

During his frequent trips to the Computer Center, Pete Owens discovered computer graphics, in particular, the ability to display three-dimensional functions with the CalComp plotter. We recognized how useful this facility would be in representing orbitals and electron density functions and published some applications to the methane molecule and methyl anion.[249] We subsequently published a book containing many such computer plots to help undergraduates understand orbitals.[250] This theme of

# A Lifetime of Synergy with Theory and Experiment

*Two theoretical chemists, John Pople (left) and Henry F. (Fritz) Schaefer, III, at dinner in 1995. John helped us get started with ab initio calculations. I have collaborated with my former Berkeley colleague, Fritz, now at the University of Georgia, in several papers.*

using ab initio computations to model chemical concepts, in part with electron density functions and frequently with the aid of computer graphics, has been an important part of my research ever since. Pete Owens is on the faculty of San Mateo College, just south of San Francisco, and has spent frequent summers leading Sierra Club trips to Nepal.

The systematic studies of Pople's group using first the minimum basis set STO-3G and then the split valence basis set 4-31G showed that the single-determinant Hartree–Fock method can give satisfactory structures for normal organic molecules and cations consisting of first-row elements.[251] For better results and especially for small ring compounds it is generally necessary to add "polarization functions" (e.g., d orbitals) to give the wave functions additional flexibility. This approach usually works quite well for normal closed-shell neutral molecules and cations in which the HOMO and LUMO are far apart. Computations with Pople's GAUSSIAN series of programs were especially convenient and gave access to quantum chemistry to many organic chemists. Early computations were often done with "standard structures" using normal or averaged bond distances and angles. The determination of structures located at local minima on the potential energy surface was a more difficult task that required much more computer time because a variation of individual structural parameters was required until the energy minimum was found. With later versions of the GAUSSIAN series and other recent programs even this problem is solved automatically for most systems by the program. Thus, the energy of a given conformation of nuclei can be obtained readily, and many energies and energy changes have been computed.[252]

An example of the kind of approach possible is demonstrated by a study in the early 1970s to evaluate the role of d orbitals in the stabilization of carbanions by adjacent sulfur. Sulfur had long been known qualitatively to stabilize adjacent lone pairs in general and carbanions in particular. The traditional explanation was in terms of multiple bonding to sulfur that required expansion of the sulfur octet via d orbitals. This concept can be expressed by resonance structures as in **21**.

$$H_2C^- - SR \longleftrightarrow H_2C = {}^-SR$$

**21**

A more common example is in the expanded sulfur resonance structures for sulfuric acid, **22**.

$$HO-\overset{\overset{O}{\|}}{\underset{\underset{O}{\|}}{S}}-OH \longleftrightarrow HO-\overset{\overset{O^-}{|}}{\underset{\underset{O}{\|}}{S^+}}-OH \longleftrightarrow HO-\overset{\overset{O^-}{|}}{\underset{\underset{O^-}{|}}{S^{++}}}-OH$$

**22** [2]

To test the importance of this concept we calculated the gas-phase proton affinity of thiomethyl anion, eq 22, with and without d orbitals.

$$HSCH_2^- + H^+ = HSCH_3 \tag{22}$$

The idea was that if d orbitals are important in stabilizing the thiomethyl anion, then a calculation in which they are omitted could not give the anion its proper stability, and the calculated proton affinity would be too high. In fact, the two calculations gave calculated proton affinities that differed by only a few tenths of a kilocalorie.[253] That is, inclusion of d orbitals in the basis set had virtually no effect on the relative stability of the carbanion. An alternative explanation for stabilization by sulfur came from a study of electron-density difference maps that show that the principal stabilizing effect is that of polarization. Second-row elements such as phosphorus and sulfur are more polarizable than first-row elements and can stabilize an adjacent charge by the polarization phenomenon of classical electrostatics. One symbolic representation of this effect is given as **23**:

$$H_2C^- — \overset{\leftarrow\rightarrow}{SR}$$

**23**

In another study showing how computations on simple model systems can provide added understanding of physical organic concepts, we examined substituted benzenes perturbed by nearby point charges and dipoles to model the effect of reactions not conjugated to the ring.[254] The results were analyzed in terms of relative inductive and polarization contributions to the corresponding Hammett substituent constants. One of my teaching philosophies is illustrated by this work. One of the graduate students involved in this study, Spiro Alexandratos, loved theoretical organic chemistry and wanted to do only computations. Because he was an organic chemist, however, I insisted that he do some experimental work, which he did, albeit reluctantly. I still believe that all organic chemistry graduate students, even those who are primarily interested in theory, should know experimental chemistry. After graduating, Spiro was employed at Rohm and Haas as a research chemist doing only experimental work. He found a special pleasure in aspects of polymer chemistry and subsequently left industry for an academic position at the University of Tennessee. He is now a Professor of Chemistry there following his own research primarily in experimental polymer chemistry!

## Electron Density Functions

Applications of electron density functions came readily to mind with my background as a physical organic chemist. The curved arrows used in the electronic theory of organic reactions represent electron density changes that should, in principle, be apparent in the electron density function. So it was that our earliest ventures into ab initio computations included the study of electron densities. An important example occurred in the early 1970s when we had occasion to study the electron density function of methyllithium. This study was an obvious outgrowth of our interest in carbanions. Because of computer limitations we were restricted to small molecules, and our experimental carbon acidity program actually involved alkali metal salts and especially organolithium compounds. Ethyllithium was used as a reagent in our earliest experimental studies as a route to lithium cyclohexylamide.

I had been taught, and I taught in my physical organic courses, that the carbon–lithium bond is a polar bond that retains a high degree of covalent character as expressed by the resonance structures, $H_3C-Li \leftrightarrow H_3C^-\ Li^+$. "Covalent" and "ionic" bonding are qualitative terms that do not correspond to quantum mechanical operators and therefore have no unique quantitative definitions. Nevertheless, they are qualitatively useful in understanding complex chemistry. For example, organolithium compounds generally dissolve in organic solvents, have appreciable volatility, and do not react as rapidly as typical ionic salts. It was thus reasonable to consider them as typical organic compounds in which the C–Li bond has many of the same properties as C–F, C–O, and other similar bonds between atoms of different electronegativity but with properties associated with covalence. Consequently, it came as quite a surprise to me to see in the electron density function of methyllithium that the amount of shared electron density between carbon and lithium is rather small compared to that between carbon and hydrogen (Figure 18).

We published[255] this result and called attention to the "absence" of covalent character in the carbon–lithium bond. Of course, we did not mean that there is no covalency in this bond—all bonds have some orbital overlap and hence some covalency—but we intended to point out that carbon–lithium bonds are more

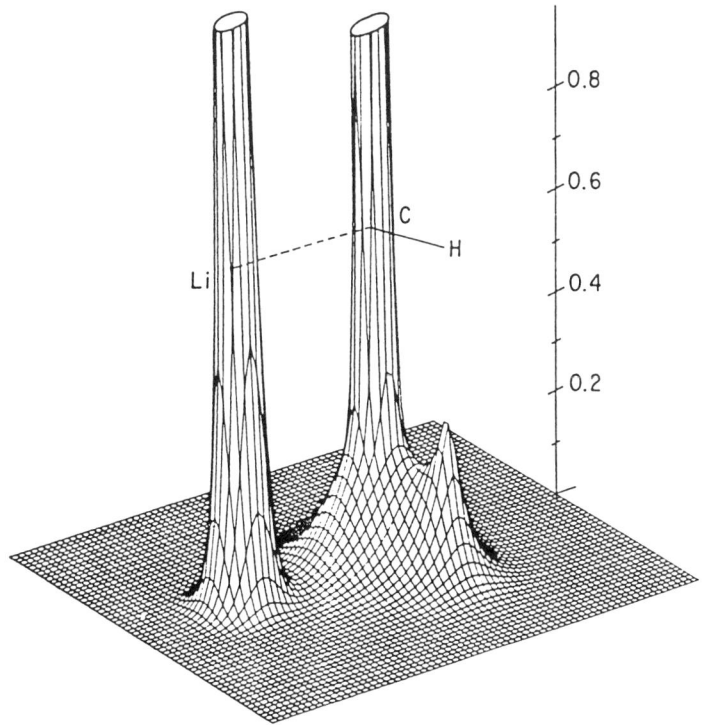

Figure 18. Electron density function for a HCLi plane of methyllithium, $H_3CLi$. The low value of the density between carbon and lithium can be compared with the high ridge of the C–H bond to show relatively low C–Li covalency. (Reproduced from reference 255. Copyright 1976 American Chemical Society.)

ionic than had been thought. In particular, many of the properties of organolithium compounds that had been considered to demonstrate covalent character do not stand up to close inspection. For example, the solubility and volatility are properties of aggregates in which the ionic centers are surrounded by hydrocarbon shells. Reactions of alkyllithium compounds are slower than typical ionic reactions because it takes work to separate ion pairs as required in most reactions. Nevertheless, our paper generated immediate controversy. I recall one referee of a NSF grant proposal, in which I proposed continuing this study of organolithium theory, who opined, "Streitwieser is surely wrong"! (The proposal was not funded.)

I had frequent debates on this question with Paul Schleyer° until he also, on the basis of his own studies, became convinced that the C–Li bond has a higher degree of ionic character than he had thought. Important parts of the reasoning involve the highly nondirectional character of bonds to lithium, the ability of pure ionic models (for example, lithium cations with no valence shell) to mimic organolithium structures, and the close comparison of relative organolithium bond strengths to the still more ionic bonds of organosodium compounds. Most experts in the field now agree[256] that the carbon–lithium bond is highly ionic but does have a small but significant amount (10–20%) of covalent character. The principal outcome of this change in interpretation is that simple ionic models suffice to explain the chemistry of organolithium compounds. It is in this sense that I believe our methyllithium paper[255] is a significant contribution to chemistry.

One example of how effective a simple ionic model can be is found in our computational study of cyclopentadienyllithium.[257,258]

From left: A. Streitwieser, Peter Stang, Kurt Mislow, and Paul Schleyer, at a Symposium held in honor of Paul's 60th birthday, Erlangen, Germany, April, 1990. Peter Stang is a former student. I have known Kurt Mislow and his stereochemical research at Princeton for many years. Paul Schleyer and I are long-time friends, collaborators, and competitors.

As expected, the lithium cation is located over the center of the five-membered ring. What was unusual was the finding that the ring hydrogens are bent away from the lithium by about 2°. This result was explained on simple electrostatic grounds. The ring charge is not divided equally between the two faces but is attracted to the lithium face by the cationic charge. The resulting rehybridization at each ring carbon then directs the hydrogens to the opposite face. This explanation was confirmed by showing that exactly the same phenomenon occurs when the lithium cation is replaced by a point positive charge containing no orbitals.

Later in the 1970s, John Collins, who was then a postdoctorate in my group, recognized how the separation of variables in the electron density function would facilitate integration. We devised[259,260] a "projection function" in which the electron density is integrated along one Cartesian axis. The result gives a projection of the entire density onto a plane, usually chosen as a convenient plane of the molecule. The result could be readily integrated numerically about regions of an atom to give an integrated population.

These populations are approximations to the spatial electron populations rigorously defined by Richard Bader°.[261,262] Bader showed that the "zero-flux" surface (on which the derivative of the electron density function normal to the surface is zero) of the electron density about each atom in a molecule has important properties. Quantum mechanical operators act within such a defined surface, the "basin", to give "atomic" properties, which are often similar in different molecules and which can be summed to give the molecular property. The total number of electrons within this surface or basin is the integrated population associated with an atom. Bader's zero-flux surfaces are curved surfaces in three-dimensional space. Our related surfaces are vertical curtains because of the integration along one axis, but the integrated projection populations are similar to Bader's, especially for cases, as in lithium compounds, in which the electron density falls to a low value between atoms. The path of maximum electron density between two nuclei is called the "bond path" by Bader, and the minimum along this path is the "bond critical point". It seems reasonable to consider that a necessary but not sufficient criterion for covalency is significant electron density at the bond critical point.

*Richard F. W. Bader giving a talk in 1990. Richard is a theoretical chemist who received his early training as a physical organic chemist. He is best known for his work on the topological properties of electron density functions. Richard spent two months at Berkeley as a Visiting Miller Professor in 1993 during which we furthered our fruitful collaboration.*

The projection function for methyllithium is shown in Figure 19. The integrated population gives lithium a charge of +0.87, a value that compares well with the value of +0.91 calculated by Bader's method.[263] Thus, although Bader's basins have a fundamental significance that the projection functions do not share, the integrated projection functions do have a qualitative resemblance. Because the projection function is much faster and easier to

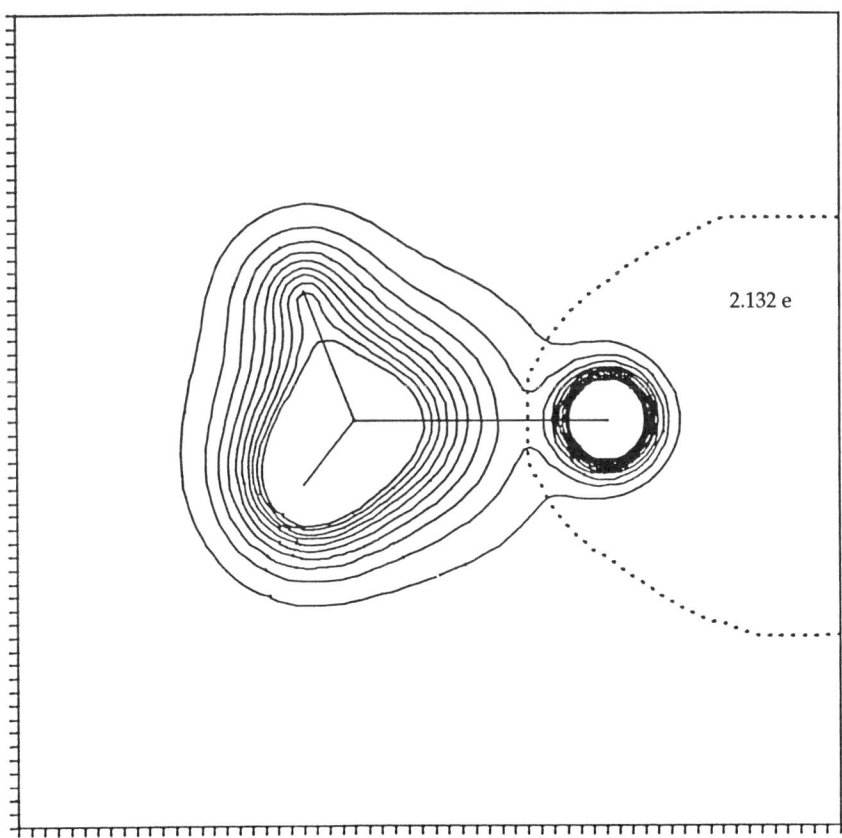

*Figure 19. Projection function for methyllithium arranged H$_3$CLi. The integrated population of lithium is 2.13e, corresponding to a charge of +0.87.*

compute than Bader's integrations, it may well continue to be useful as a tool in the interpretation of electronic structure, especially because it also lends itself well to graphical depiction for many types of systems. The original program, PROJ, for calculating the projection function has some flaws and has now been completely rewritten as PRODEN and is available from QCPE.[264,265]

David Grier was another graduate student who combined experimental work on carbon acidity, particularly on some sulfur compounds, with extensive theoretical calculations. Because of his interest in stabilization of carbanions by sulfur, he studied the application of the projection function approach as in the following instructive example.[266] Consider the "projection difference" function for the deprotonation of methanethiol to thiomethyl anion, eq 23.

$$HSCH_3 \rightarrow HSCH_2^- \tag{23}$$

The two compounds are isoelectronic, and one can readily calculate the projection functions of both compounds and subtract them. A meaningful difference function is only obtained if the nuclei are kept in the same locations except for the deleted proton. The structures actually do differ somewhat, but the effect of this change is generally small. The difference function $P(HSCH_2^-) - P(HSCH_3)$ for the HSC–H* plane (where H* denotes the proton removed) is shown in Figure 20. In this figure the proton removed is on the lower left and the HS bond is on the right side. The large negative contours show the electrons held close to the methyl hydrogen in the mercaptan that are lost and pulled toward the carbon in the carbanion. There is a resulting polarization at the sulfur and particularly along the HS bond.

This figure can be compared with a similar one for the deprotonation of ethane to give ethyl anion. Figure 21 shows the corresponding projection difference plot, $P(^-CH_2CH_3) - P(CH_3CH_3)$. The two figures are qualitatively similar, the principal difference being that a methyl group is somewhat less polarizable than a SH group. Especially instructive is Figure 22, which shows the corresponding difference plot for deprotonation of methane with an argon atom placed where the sulfur would be in methanethiol. The argon atom serves as a polarizable group not bonded to the carbon acid. Yet, the polarization in the argon atom (shown as an asterisk in Figure 22) is qualitatively similar to that of the SH group in Figure 20 and the $CH_3$ group in Figure 21. The principal difference is that argon is spherically symmetric and the polarizability of the other groups is anisotropic.

A corresponding difference plot of the electron density itself in any given plane is not meaningful because the number of electrons may change; that is, from one compound to the next, electron density can shift from one plane elsewhere. In the projection plot the total number of electrons remains the same for both species, and the integrations in Figures 20–22 must sum to zero.

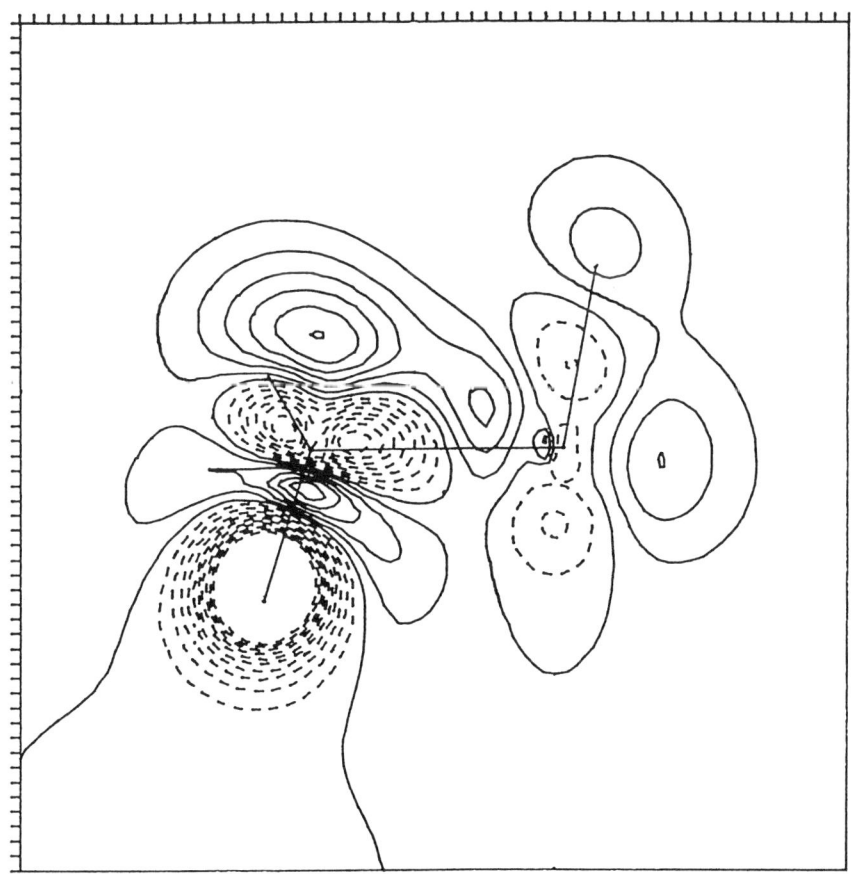

*Figure 20. Projection function difference plot $P(^-CH_2\text{-}SH) - P(CH_3\text{-}SH)$. Contour levels are 0.01 e au$^{-2}$, and the first solid contour is the zero contour. The proton removed is on the lower left; the SH proton is on the upper right.*

The effect of a lone pair is often represented in MO theories by its interaction with an appropriate σ* orbital,[267] an example of anionic hyperconjugation (*see* previous section titled Organofluorine Carbanions); in thiomethyl anion, this interaction is with the S–H σ* orbital and is equivalent to polarization along this bond. The argon example in Figure 22, however, shows that a comparable type of polarization occurs even when there is no bond! Finally, these examples show further that d orbitals play no special role in the stabilization by sulfur.

Atom populations and charges are often given as Mulliken

*Figure 21. Projection function difference plot $P(^-CH_2-CH_3) - P(CH_3-CH_3)$. Contour levels are 0.01 e au$^{-2}$, and the first solid contour is the zero contour. The proton removed is on the lower left; the staggered $CH_3$ group is on the right.*

populations that are defined on the basis of the molecular orbital coefficients and basis functions.[252] That is, the Mulliken populations are basis set populations in which the electrons contained in a given basis function are assigned to the atom on which the function is centered, regardless of the spatial extent of the function. Mulliken populations for lithium in methyllithium give much less cationic character to the metal, typically about +0.5 to +0.6,[255] and more in line with conventional thought. The explanation comes from the basis set character of Mulliken populations in the following way. Lithium and carbon are generally given the

# A Lifetime of Synergy with Theory and Experiment 163

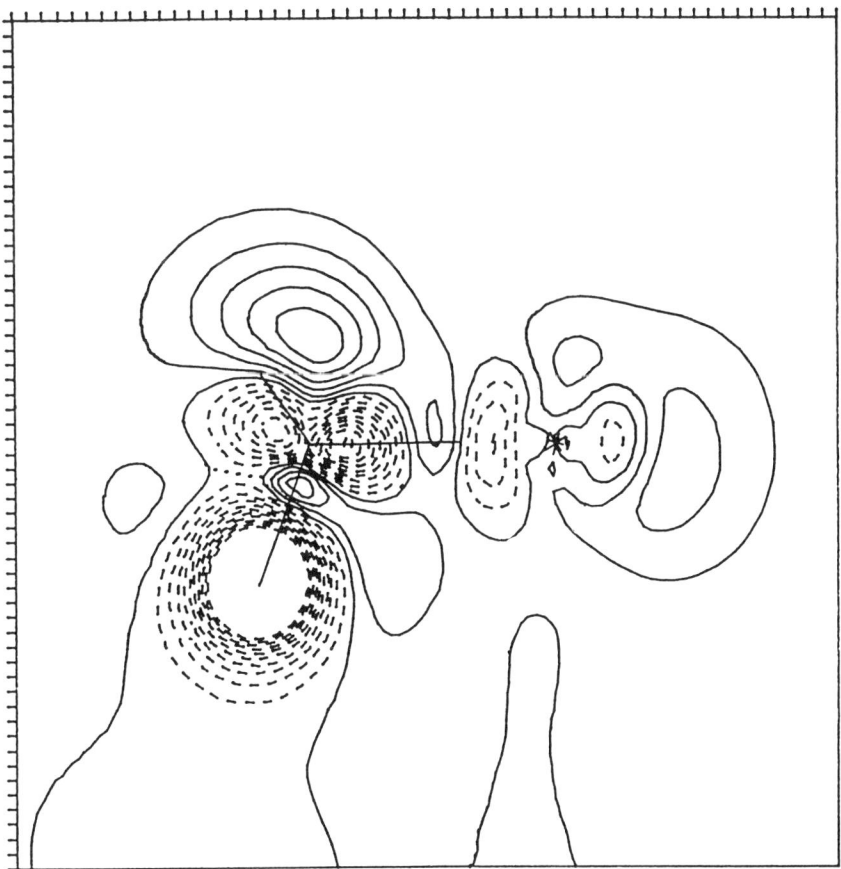

*Figure 22. Projection function difference plot P($^-$CH$_3$···Ar) – P(CH$_4$···Ar). Contour levels are 0.01 e au$^{-2}$, and the first solid contour is the zero contour. The proton removed is on the lower left; the argon atom (asterisk) is on the right.*

same number of basis functions representing the 1s, 2s, and 2p atomic orbitals of these first-row elements; however, lithium cation has only two electrons to be described by these functions, whereas a carbanion is much more electron-rich. Thus, electrons close to carbon will nevertheless "use" diffuse functions centered on lithium for their description and will be assigned by the Mulliken procedure to lithium. This artifact in Mulliken populations will occur generally whenever an electron-poor, function-rich center is close to an electron-rich, function-poor center. This type of result emphasizes that even modern quantum chemical

**24**

results require care and understanding in their interpretation. The point was brought home to me in calculations of 1,3-dilithiopropene.[268] In the most stable structure, **24**, both lithiums are π-bonded to a propylidene dianion[269] and the resulting structure is clearly related to the triple ions discussed previously. In further calculations to try to understand the resulting wave function, I realized that the lithium orbitals were taking some of the mathematical role of the diffuse carbon orbitals of the parent carbanion. This type of "basis set superposition error" results when basis sets are too small or unbalanced. Similar superposition errors were analyzed in calculations of ethynyllithium.[270]

I also like to emphasize one other important point about the controversy between Paul Schleyer and me concerning carbon–lithium bonding. This controversy was expressed in publications and at chemical meetings and seminars but was always maintained on a scientific level. Our differences of opinion never became acrimonious and did not affect our friendship. In fact, during the past decade we collaborated with the help of a NATO grant and have a number of joint publications.

Much of our recent work has made use of projection functions and integrated populations. Boris Kohler was a remarkable Swiss graduate student with a Ph.D. in mathematics who pursued his more recent interest in photochemistry by experimental work with Professor Dauben and theoretical work with me. He modified the PROJ program to work with configuration interaction wave functions and computed the projection function differences in the $n \rightarrow \pi^*$ transition of formaldehyde. His results showed some differences from conventional thought that we rationalized by the dominating effect of oxygen electronegativity.[271] Together with a postdoctoral student, Birgitte Schilling, Boris made comparisons of carbocations with isoelectronic boron compounds that showed much more sigma polarization in the cations than expected by accepted considerations of resonance structures. Because the results were only at a minimum basis set level, they were never published although we did prepare a rough draft of a manuscript. The conclusions were later confirmed by Bader in related studies of integrated populations using larger basis sets.[272] Polarizations of sigma electrons in substituted benzenes were also demonstrated by Eric Vorpagel[273] and in carbonyl groups by David Grier.[274]

Dan Bors was primarily an experimental graduate student who helped initiate the cesium ion pair acidity work in THF (*see* previous section titled Carbanions in Tetrahydrofuran), but he also did some theoretical work for his dissertation. He showed that the sulfonyl SO bond in sulfones and derived carbanions is highly polar with a large negative charge on the sulfonyl oxygens;[275] that is, the Lewis structure with a single $S^+$–$O^-$ bond dominates the electronic structure of sulfoxides and sulfones compared to the "expanded octet" S=O structure in **25**.

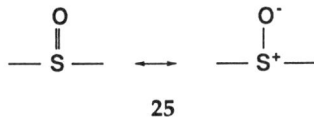

**25**

As a result of the highly negative oxygens, a significant calculated structure for the lithium salt of dimethyl sulfone is one in which the lithium is bridged between the two sulfone oxygens and far from the carbanion carbon, a type of structure shown by X-ray crystal structures shortly thereafter. Similar findings were made for some bonds in phosphorus compounds. Extensive calculations about the same time by Bob McDowell showed that the PO and P–CH₂ bonds in phosphine oxide and methylenephosphorane, respectively, are also highly polar with little multiple-bond character.[276,277] Andrzej Rajca started his Miller postdoctoral fellowship in 1985 in part to study the "aromatic" character of various threefold symmetric species related to trimethylenemethane dianion and found instead that the bonding in these compounds is generally highly polar; that is, ionic interactions rather than conjugation dominate the electronic structure.[276–279] Scott Gronert also did both experimental and theoretical work for his dissertation; he and Rainer Glaser found that bonds to silicon are often quite highly polar.[280]

Many of these studies were facilitated by having our own computer. Wilhelm Maier, who was then an Assistant Professor at Berkeley, and I had applied successfully to NSF for an equipment grant to enable us to buy a VAX 750 computer system. Having this computer in our groups made a major difference in our ability to do significant computations. For example, my rate of publication of theory papers doubled after we obtained this system. The VAX is now an outmoded computer, and it has been replaced by a new generation of remarkable "workstations". These small but powerful computers have allowed us to continue with theoretical studies.

## Carboxylate Resonance

With these varied studies showing a role of ionic effects and polarizations not revealed in normal chemical symbols or in resonance structures in common use, the stage was set for another dramatic development. This story started in June, 1984, when I visited Oregon State University to give a seminar and to discuss research with chemistry faculty. Professor T. Darrah Thomas, whom I had known since he was a graduate student years earlier in Berkeley, and his student, Michele Siggel, told me of their interpretation of X-ray photoelectron spectroscopic results of acetic acid and ethanol and related compounds. Siggel and Thomas had concluded that the higher acidity of the carboxylic acid was already inherent in the potential at the acidic proton. That is, the higher acidity of acetic acid compared to ethanol is implied in the electronic structure of the neutral acid, and the additional resonance energy of carboxylate ion is relatively unimportant.[281] It took a bit of explaining to get me to understand the reasoning behind their conclusions. After all, these ideas give only a minor role to the resonance energy of carboxylate anions and differ radically from conventional organic chemical thinking as given, for example, in many organic textbooks including my own!

Darrah had ample opportunity to educate me because we spent the last couple of days of my visit fly-fishing. Oregon is great country for this sport, and this was our first opportunity to

## A Lifetime of Synergy with Theory and Experiment

go flyfishing together. Except for my summers in Idaho, I have few other opportunities to flyfish, and I greatly enjoyed the extra few days on the Deschutes River with Darrah. We talked about the interpretation of the new experimental results and made plans to compare them with theoretical work.

Several weeks later Michele came to Berkeley and spent a few days with my group. She learned to use the ab initio programs on our VAX 750 system and completed several calculations. These were used in her dissertation for the computation of potentials at hydrogens and confirmed her experimental conclusions. After finishing her Ph.D., Michele Siggel came to Berkeley in January, 1987, in part to continue further work in photoelectron spectroscopy and in part to do further calculations in my group with our VAX system. I did not have funds available at the time to provide a postdoctoral stipend, but I was able to arrange for her to support herself as a temporary lecturer in our freshmen chemistry program.

During Michele's first visit, Steve Bachrach in my group was just in the process of showing that the integrated populations of oxygen in several types of compounds differ but little.[282] In her own calculations of formic acid and formate ion, ethanol, and ethoxide ion, Michele showed[283a] that the oxygens also have high negative charges that do not differ much in these different functional groups. That is, deprotonation of a carboxylic acid is not accompanied by much of an increase in the electron population of the carbonyl oxygen because that population is already so high. Instead, polarization effects distribute much of the electron population associated with the departing proton to other parts of the molecule. Alternatively, the results can be viewed in terms of the inductive effect of the carbonyl group because of its high $C^+-O^-$ polarization.[283b] The argument may be clarified by consideration of resonance structures of a carboxylate anion, **26**. Structures **26a** and **26b** are the conventional structures that show delocalization of charge from the original hydroxy oxygen to the carbonyl oxygen. Structure **26c** may be called the polarization structure, in

26a        26b        26c

which the additional charge on the original hydroxy oxygen polarizes the carbonyl group. Even these structures are inadequate symbols because the bonds to oxygen shown as simple lines have additional polarizations toward oxygen. All three structures contribute to the resonance energy of the carboxylate anion, but **26c** is primarily responsible for the inductive effect.

$$
\underset{\textbf{27a}}{R-C{\overset{CH_2}{\underset{O^-}{\phantom{|}}}}} \leftrightarrow \underset{\textbf{27b}}{R-C{\overset{CH_2^-}{\underset{O}{\phantom{|}}}}} \leftrightarrow \underset{\textbf{27c}}{R-\overset{+}{C}{\overset{CH_2^-}{\underset{O^-}{\phantom{|}}}}}
$$

Similarly, structures **27a** and **27b** are the delocalization structures for the enolate ion of a methyl ketone, and **27c** is the polarization structure; the additional charge on the enolate oxygen polarizes the double bond in the manner shown. This structure is apparently more important than generally recognized. Once again, all three structures contribute to the resonance energy. This point of view emphasizes the significant difference between delocalization and resonance. It also emphasizes the important role of classical polarization. The important role of polarization was also dramatically demonstrated by the gas-phase acidity studies of Brauman.[284]

Meanwhile, through the generosity of Professor Bader and the help of Professor Wiberg, we obtained a copy of Bader's programs suitable for use on our VAX computer system. The integrated populations using this program confirmed the general results obtained by the projected densities.[283b] During this period, Bader's group[285,286] showed the importance of local polarizations in charge distributions and Wiberg and Laidig[287,288] independently made observations similar to ours concerning the importance of polarization relative to traditional conjugation in interactions with carbonyl groups. The same principles were extended to nitrous and nitric acids.[289,290] It appears that the high positive charge on nitrogen rather than resonance in the anion is responsible for the high acidities of these acids. Similar results were found recently for the effect of oxygen in dimethyl sulfoxide and dimethyl sulfone. The increased acidities of these compounds compared to dimethyl sulfide are already manifest in the acid; electronic reorganization in the anion plays a relatively small role.[291,292]

These types of studies are still in progress, and their ultimate impact on our thinking about electronic structure and bonding remains to be assessed; however, the main outlines of what will probably emerge seem fairly clear. Many of the phenomena previously rationalized by resonance structures involving conjugation and charge transfer (the "curved arrows") will instead be shown to involve polarization of electron density in an almost classical sense.

I expect this to be especially true of allylic systems containing heteroatoms. Allylic resonance structures and "curved arrows" have been important in chemical thinking to show polarization effects primarily in π-electronic systems. Polarization of sigma bonds is not well depicted in our usual chemical symbolism and has therefore tended not to get the attention it deserves. In conjugated systems containing highly electronegative atoms such as oxygen, conjugation effects of the type normally expected from π-MOs or resonance structures abbreviated with the curved arrow symbolism may well still be significant but are offset by compensating sigma back-polarization such that the net population of the atom has changed but little. That is, the net charge transfer implied by such conjugation and by simple resonance structures may be much smaller than expected; however, the changing roles of sigma and pi density imply that the quadrupole moment around the atom will have changed.

How these concepts affect theories of reactivity and our interpretation and use of chemical symbols are parts of questions that can only await further study. Such research will largely, I suspect, be theoretical in nature, and organic chemists who are familiar with conventional symbolism will undoubtedly be important contributors. But vital contributions will certainly also come from experimental results involving imaginative use of new spectroscopic tools such as in the original Siggel and Thomas work. The result may be a much more classical picture of electronic structures making use of charges, dipoles, polarizability, and induced dipoles, a picture that emphasizes the physics of the classical electrostatics of electron distributions.

These examples point up how modern computational quantum chemistry can affect our perception of qualitative chemical concepts. It also emphasizes what I believe is one of the important uses of modern computational quantum chemistry. In principle, one can now do calculations that reproduce experimental numbers with a useful degree of precision. In many cases, however, merely reproducing experiment does not teach us much unless the results are analyzed to provide understanding of the important contributions to the numbers in terms of reference systems and simpler models. Classical models in particular have always been important in chemistry, and one of the things that quantum calculations can do is to tell us which classical effects are important. For the sulfur-stabilized carbanions already discussed, for example, this substituent effect can be readily understood in terms of classical polarization in which the polarizability along a bond is greater than the transverse polarizability perpendicular to the bond as shown by Denbigh in 1940;[293] that is, MO theory provides the mathematical implementation of classical polarizability. I find it especially amusing that sophisticated modern computations end up with a simple model that would have been completely understandable by the chemists of more than a half-century ago!

## Transition States and Ion-Pair Reactions

Until recently, computer limitations restricted computations to relatively small molecules and to only a few simple reaction transition states. With the advent of supercomputers and inexpensive minicomputers and workstations, quantum mechanical calculations are already having an increasingly important role in chemistry with applications to large systems closer to reality. Computations have been important in the past in determining structures of carbocations, and we are already seeing more applications to other reactive intermediates important to synthetic chemists. Examples are organometallic compounds and the role of aggregation, effects of solvation, and conformational effects.

Computations of transition states are increasingly common in the modern literature. Many of these calculations involve ions and therefore have reduced relevance to reactions in solution where solvation of ions is so important. We have tried to minimize this limitation by emphasizing reactions of ion pairs because such reactions are so common in synthetic chemistry. Indeed, many reactions in nonpolar solvents written by chemists as reactions of ions in fact are reactions of ion pairs. Moreover, the study of ion pairs is also closely related to our experimental studies of carbon acidity in nonpolar solvents. One of our first such calculations was the reaction of lithium hydride with formaldehyde.[294] The transition state is highly bent, with the lithium cation coordinated to both the reacting hydride ion and the carbonyl oxygen. This structure was also found by the Schleyer and Houk groups[295], and the general motif of highly bent transition states has turned out to be common for ion-pair reactions.

Proton transfer reactions between methyllithium and methane and between lithium amide and ammonia also involve highly bent transition states.[296,297] We studied the reaction of lithium amide with methane and the electron density changes during reaction[298] as the first part of a more extended study with other hydrocarbons to model our reactions in cyclohexylamine.[299] The ultimate goal is to compare the calculated activation energies with experimental rates of reaction of the hydrocarbons with cesium cyclohexylamide and to be able to convert the kinetic acidities of weak acids to equivalent equilibrium acidities. Of course, an amide may not be an unreasonable model for a cyclohexylamide, but lithium cation is not a good model for cesium. Accordingly, calculations currently in progress involve cesium cation using a pseudopotential model for this large system.[300]

One interesting result involves the effect of the cation on the transition state structure. The transition state for the reaction of lithium amide with methane is shown in Figure 23;[298] the N–H–C bond angle is 163.2°. For the corresponding ionic transition state, with no gegenion, N–H–C is almost linear, with a bond angle of 178.6°. If the lithium cation is solvated with a molecule of ammonia, the bond angle increases to 164.8°, and if a sodium cation is used, the bond angle is 169.0°. The corresponding cesium transition state extends this pattern.

Rainer Glaser was an unusually versatile and productive student from Germany who spent a year at Berkeley taking courses and doing experimental work in uranocene chemistry and theoretical calculations on electron density functions. The theory work formed the basis for his Master's dissertation at Berkeley, and the experimental work was used for his *Diplom* thesis at the University of Tübingen. After completing his examinations in

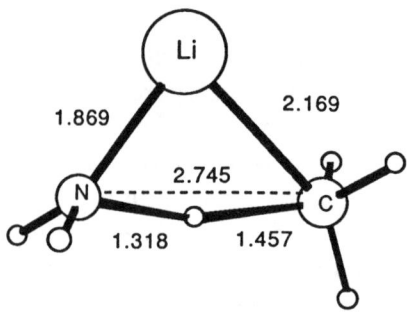

*Fig. 23. Transition state (6–31G\* basis set) for the proton exchange of methane with lithium amide. Distances shown are in angstroms.*

Tübingen he returned to Berkeley for his Ph.D. His reports and first drafts of papers were as good as those of American students, but he worked continually to improve his already excellent English. In the course of trying to answer his questions I had often to explore why some expressions "sounded right" and others did not; his questions helped me to understand my own language better!

Among the calculations Glaser did during this period in the mid-1980s were extensive studies of lithium and sodium salts of oximes and enol ethers. The most stable lithium carbanion salt of acetaldoxime, for example, was found to be the structure in Figure 24.[301–303] The lithium is coordinated to the nitrogen, oxygen, and carbanion carbon, but this coordination site is also indicated simply from the electrostatic potential of the carbanion. The *syn*-orientation of the NOR group relative to the carbanion carbon also explains the observed *syn*-stereochemistry in alkylation reactions of oxyimine enolate salts. Further computational studies of isomerizations and racemizations within such salts pointed up interesting questions of stereochemistry.[303] Rainer found, for example that the transition state for the coordination change of lithium from one face of the carbanion to the other, a conversion of one enantiomer to the other, does not go through an achiral transition state. In this compound the ONCH=CH$_2$ group defines a plane. As shown in Figure 24, the coordinating lithium is on one side of this plane and the hydroxy hydrogen is on the other. In passing from one enantiomer to the other, both the lithium and the hydrogen must pass through this plane, *but they do so at*

*A Lifetime of Synergy with Theory and Experiment* 173

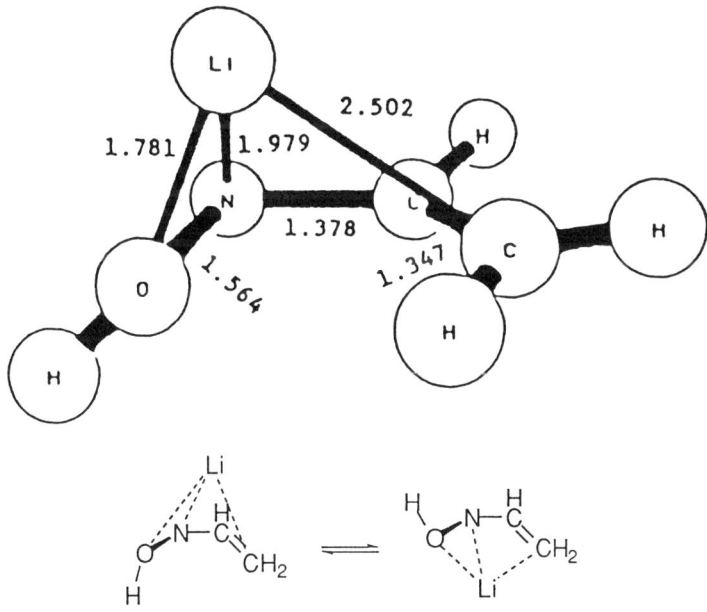

Figure 24. *Movement of lithium from the top face to the bottom converts one enantiomer to the other.*

*different times!* The system remains chiral throughout, and two enantiomeric transition states are involved in the reaction! Mislow[304] had pointed out earlier the possibility of totally chiral pathways to racemization, but such reactions appear to be rather rare.

Many calculations have been made of $S_N2$ reaction transition states, frequently of the simplest example, $F^- + CH_3F$. The transition state structures generally show a linear arrangement of nucleophile, central carbon, and leaving group. In an early such calculation on this reaction, Duke and Bader[305] found high integrated electron populations about the two fluorines, giving each a charge of $-0.86$; the $CH_3$ group was found to be $+0.73$. This calculation gives the central carbon much more carbocation character than is generally thought for $S_N2$ reactions. Similar results, however, have been obtained by Shi and Boyd[306,307] for a number of $S_N2$ reactions on methyl compounds with different nucleophilic and leaving groups. Charges on the methyl group are about $+0.6$, and the electron densities at the bond critical points between carbon and entering or leaving groups at the transition state are rather low. These results change somewhat with incorporation of configuration interaction corrections[308] but still indicate that the resonance structure C contributes quite substantially to the resonance hybrid 28.

$$[ Y-CH_3 \quad X \leftrightarrow Y \quad CH_3-X \leftrightarrow Y^- \quad CH_3^+ \quad X^-]$$
$$\phantom{[\,} A \phantom{XXXXXXX} B \phantom{XXXXXXX} C$$

28

We have recently been doing calculations of ion pair $S_N2$ transition states with interesting results.[309] The American chemist S. F. Acree had shown[310] in the early years of this century that undissociated ion pairs, although generally less reactive than free ions, are involved in many displacement reactions.[310] Our calculations of such reactions found the same motif as with other types of ion-pair reactions involving transition states with no net charge, namely kite-shaped structures such as that between LiF and $CH_3F$, 29.

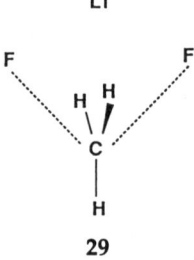

29

The F–C–F bond angle is 80°. Thus, the $S_N2$ transition state does resemble a triple ion, with the fluorides bent toward the lithium cation in the corresponding ion-pair transition state. With a larger cation or larger entering and leaving groups, the bond angle increases. At the same basis set level, for example, the bond angle is 92.5° for the reaction of LiCl with $CH_3Cl$, and for sodium cation the angles are larger still. This whole behavior is that expected for a collection of ions!

But, we may ask, if the methyl group in an $S_N2$ transition state approximates a methyl cation, why do we not see normal carbonium ion type or $S_N1$ substituent effects? We could explain the lower reactivity of secondary carbons by steric effects, but allylic and benzylic systems show different substituent effect reactivity patterns in typical displacement reactions compared to

# A Lifetime of Synergy with Theory and Experiment

solvolyses and other carbonium ion reactions. The answer may come from similar calculations on allylic systems.[311] The transition state for the reaction of F⁻ with allyl fluoride involves almost linear F–C–F bonding similar to the reaction with methyl. The allylic bond lengths do not show such allylic delocalization; at 1.320 and 1.480 Å, they are almost pure double and single bonds, respectively, and are virtually the same as in allyl fluoride itself. With chloride entering and leaving groups, the bonds show small changes in the direction of delocalization, 1.322 and 1.463 Å. In the reaction of allyl fluoride with LiF, however, the ion-pair transition state now shows C–C bond lengths of 1.344 and 1.413 Å; greater allylic conjugation is now apparent. These results can be interpreted with the help of Figure 25.

In the ionic transition state, allyl resonance, which delocalizes positive charge to the bottom carbon, is electrostatically impeded by the two negative fluorides. With the larger chloride ions, the positive charge is farther from the negative charges, and delocalization is more favorable. In the corresponding lithium ion pair transition state the presence of the positive lithium cation

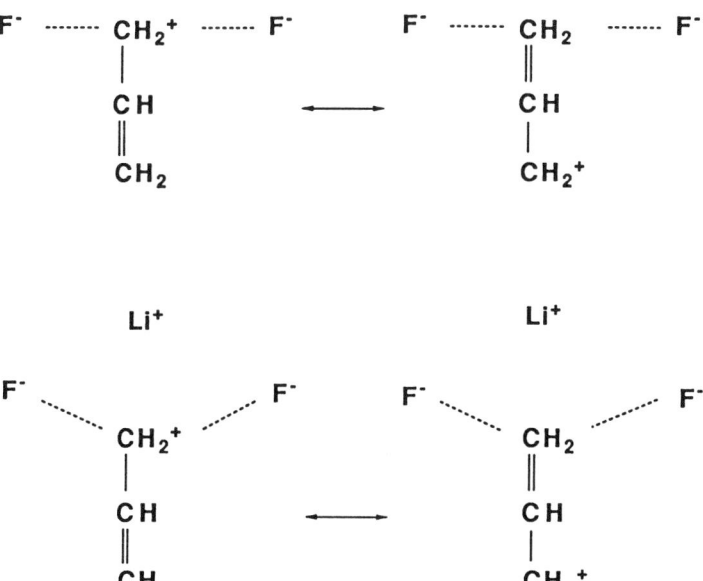

Figure 25. Resonance structures for ionic and ion-pair transition states for displacements on allylic fluorides.

facilitates delocalization even more. As expected by this model, this effect in the ion pair is reduced with a larger cation or by larger nucleophiles.

These results bring to mind an interesting result by a French group who compared Hammett σρ correlations for the reactions of substituted benzyl chlorides with ethoxide ion and with sodium ethoxide ion pairs.[312] They found ρ = +2.2 for the ionic reaction and ρ = −0.6 for the ion-pair process. The strong stabilization of the ionic reaction by electron-withdrawing groups is understandable for a reaction with a negatively charged transition state. The significant stabilization of the ion-pair process by electron-donating groups indicates that delocalization of a benzyl cation is facilitated by the positive gegenion exactly as suggested by Figure 25. This contrast between ionic and ion-pair $S_N2$ reactions has obvious corollaries and synthetic applications which, so far as I know, have not yet been recognized. As a result, we have now started to do experimental studies of ion-pair $S_N2$ reactions. I have come full circle; at the start of my career four decades ago I studied displacement reactions, and now as a result of new concepts I'm going back to the same subject!

## Future Trends

I have no doubt that further computations of structures of transition states will become increasingly important, especially as solvent effects become better treated. These computations will provide a degree of quantitative chemistry not possible by direct experiment. But they will also sharpen our qualitative concepts and the simple models required to understand and predict the chemistry of the great bulk of systems that remain too large for such computations. Symbols such as resonance structures have had an essential role in the systematic development of chemistry and will undoubtedly continue to be important in our understanding of structures and reactions.

The important trends can be seen quite clearly in the physical organic chemistry course that I taught to seniors and first-year graduate students for more than 30 years. This course was taught first by Gerald Branch and then for short periods by Melvin

Calvin and by Don Noyce before I inherited it in the late 1950s. More than 1000 students have taken this course with me, known for most of this period as Chemistry 127. The course originally made use almost entirely of Lewis structures as resonance structures and their contributions to a resonance hybrid. I introduced HMO theory and the $4n + 2$ rule as an adjunct to resonance structures. The introduction of the Woodward–Hoffmann rules required the consideration of three-dimensional MOs and their nodal characteristics, an approach that I still find pedagogically more satisfactory than the simple use of symmetry alone.

The growing application of ab initio computations to organic chemistry required the introduction of that topic and basis sets to the course together with some discussion of semiempirical methods such as Dewar's Modified Neglect of Differential Overlap (MNDO) method and molecular mechanics. Many topics once taught in this course, such as dipole moments, hybridized orbitals, and NMR, are now taught instead in the first-year organic course. Yet, Lewis structures, their use as resonance structures, and their interpretation in resonance hybrids are still important in the corresponding course today.

Lewis structures still provide an accurate accounting for electrons and a sharpness of understanding that no application of "dotted lines" can match, important as the dotted line symbolism is in conveying the blended nature of resonance hybrids. Accordingly, I still regard the proper writing and manipulation of Lewis structure symbols as one of the most important tools the budding chemist should learn early in college training. And I decry the almost universal use by inorganic chemists of the line to indicate all sorts of bonds. It is difficult if not impossible to get an accurate electron count in many of the structural symbols now in common use, whereas it is so easy with normal Lewis structures, particularly with the use of an arrow to indicate a dative bond. Unfortunately, only a few symbol elements are in common use (e.g., the line, the dot, the arrow, and the dotted line), and their mixed use has resulted in the current babel of structural symbols that lack pedagogical understanding and precision in manipulation.

Despite the complexity and size of modern MO computations, and perhaps because of them, resonance structures and other symbols of electronic structure will continue to be used routinely by organic chemists. To this end, the analysis of electronic structures using Bader's "atoms in molecule" approach and variations such as the projection function will remain important to calibrate such resonance structures. The development of concepts to help understand complex and unwieldy wave functions will continue to be a major enterprise. "Atomic charges" will continue to be a useful concept even though we know that no set of point charges centered on the nuclei of a molecule can give an adequate representation of the electrostatic potential close to the molecule. The electron density about a nucleus, and particularly the valence density, is not spherically symmetric; zero-flux surfaces are quite irregular and the "center of gravity" of the electron density is not generally located at the nucleus. Instead of the multipole expansion actually required in the basin of an atom, demand will continue for simple "charges" that give a reasonable best fit, preferably based on orbital occupancies for simplicity in computation.[313] We have seen that Mulliken populations are simply too crude for most work and really should never be used. The best such populations at the present time are probably the "natural population analyses" (NPA) of Reed and Weinhold.[314] They are straightforward to calculate and are relatively basis set independent. We have started using them ourselves, sometimes in comparison with Bader's method.[282] A related concept is that of the "natural bond orbitals" (NBO), a wave function analysis that generates a set of orbitals that are then decomposed into atomic contributions, the natural hybrid orbitals.[315] We made use of this approach in a recent analysis of the Mills–Nixon effect in cyclobutenobenzenes.[316] Eric Glendening was a recent postdoctoral coworker who came from Frank Weinhold's group in Wisconsin. While with me, he developed a "natural energy decomposition analysis" (NEDA) that uses the NBO approach to partition the interaction energy of two molecular fragments into electrostatic, charge transfer, and deformation components.[317] The NEDA procedure has some advantages over other partitioning procedures.

Thus, ab initio calculations of structures and wave functions, and their analysis by various tools, such as those mentioned as well as others not yet imagined, will be increasingly important in organic chemistry. Nevertheless, it remains true that such calculations can only be applied to molecules of limited size. Large

*When I came to one of my group meetings in 1988, something looked strange. Everyone was wearing the same T-shirt! The group designed this "Andyland" T-shirt and presented me with one. Of course, I promptly removed my shirt then and there and put on my new T-shirt. The logo is taken from an electron density function.*

molecules such as polymers, proteins, and polysaccharides will require some sort of semiempirical procedure for the foreseeable future. Molecular mechanics approaches are now indispensable for computations of conformations of such compounds and combined with computer graphics modeling and manipulation can only become increasingly important to the practicing chemist. Yet, even here ab initio computations have an important role that will certainly grow in importance. The semiempirical methods require parameters of many different types, and even now such parameters are being obtained from ab initio calculations of model systems.[318] Thus, the calculations of energy, structure, and electronic structure now done on relatively small compounds will help us to understand the chemistry of DNA spirals and protein coils in the future. I have not been involved directly in such research and have no current plans to do so, but this is an exciting prospect for the future and who knows?

# *f*-Orbital Organometallic Chemistry

## Uranocene

The 1950s were exciting years for the chemistry related to the Hückel $4n + 2$ rule. I had learned about this generalization while a graduate student at Columbia. In one of our group meetings early in 1951 Doering showed us the energy level diagram given by Hückel calculations on monocyclic π-systems in the graphical depiction later published by Frost and Musulin.[319] This discussion was my first introduction to the Hückel bond integral, β. I don't recall where Doering learned of this approach, but I imagine he was told it by his theoretical colleague, George Kimball. Doering was a major player in the experimental development of the $4n + 2$ generalization with his work on the tropylium ion, **4**.[320] A few years later, Ronald Breslow[321,322] published his clever synthesis of the amazingly stable triphenylcyclopropenium ion. Franz Sondheimer and his group[323–326] synthesized a series of large annulenes whose stabilities parallel the $4n + 2$ rule. And there were many others in this fertile period.

Among carbanions, Erich Hückel realized in the 1930s that the known relatively high acidity of cyclopentadiene is to be associated with the six π-electrons of the anion.[327] Erich Hückel was a physicist who worked between two worlds. Because he was a physicist, organic chemists paid no attention to him, and because he worked in chemistry, physicists paid him no heed. Not until the 1940s did his work have an impact on chemists. I had the good fortune to meet him during a Sabbatical leave in Germany in

1976. During a visit to the University of Marburg, Professor Karl Dimroth arranged an interview with Professor Hückel, who was then 79. I was delighted to receive a copy of Hückel's autobiography,[328] which he kindly (and overgenerously) inscribed (in English), "I present this book to Andrew Streitwieser, Jr., the most successful scientific promoter of the HMO theory." Erich Hückel died only a few years later, in 1980.

Except for writing about the $4n + 2$ rule in my MO book, I made no direct contribution to advancing its chemistry in those early years. We did later determine a more accurate value for the pK of cyclopentadiene in water,[329] but the value of 16.0 only confirmed what had been known qualitatively since the days of Thiele almost a century ago.[330] In my reading related to MO theory I noticed a 1949 report that, at a dropping mercury electrode, cyclooctatetraene (COT) undergoes a facile two-electron reversible reduction.[331] The reversibility of the reduction means that the dianion must be relatively stable. Not long thereafter the dianion was isolated as a salt by Tom Katz.[332] Our experiments with cyclooctatetraene derivatives started in the 1950s with attempts to synthesize octalene, **30**, the cyclooctatetraene analog of naphthalene.

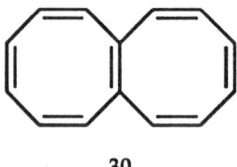

**30**

The pioneering synthetic work of Sam Andreades was ably continued by Shuji Ozawa, a versatile and clever student whose work resulted in several new cyclooctatetraene derivatives; however, we never succeeded in preparing octalene. Nevertheless, this work had a totally unexpected outcome by preparing the way for my entry into the virtually unexplored field of f-element organometallic chemistry.

In the late summer of 1968, ferrocene had long been known as had the late William Moffitt's beautiful interpretation of its ring–metal bonding in terms of the symmetry-allowed interaction of the $e_1$ π-MOs of the two cyclopentadienyls with appropriate d orbitals on the central ion.[333] A seminar on the subject at Berkeley

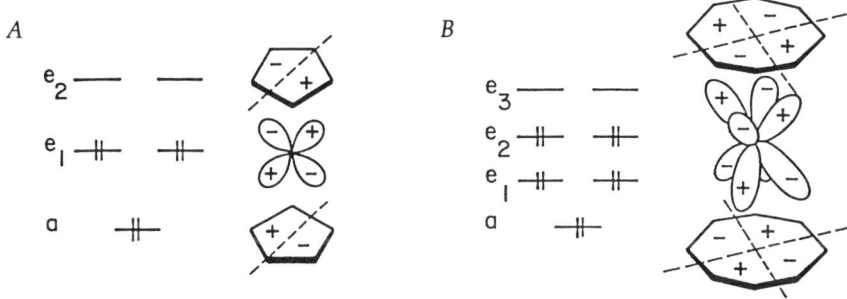

Figure 26. HOMO–LUMO interactions in metallocenes. A. The ligand π-MOs of two cyclopentadienyl anions and corresponding d orbitals of $Fe^{2+}$ of $e_{1g}$ symmetry have one shared nodal plane. B. The HOMO π-MOs of two cyclooctatetraene dianions and $f_{z2}$ orbitals of a central $U^{4+}$ of $e_{2u}$ symmetry have two shared nodal planes.

had caused me to muse one evening about this bonding interpretation. With all of our recent experimental work with cyclooctatetraene, I chanced to wonder about the possibility of a similar type of bonding with the next higher Hückel $4n + 2$ ring system with 10 π-electrons instead of 6. This thought process was clearly stimulated by our extensive use of cyclooctatetraene dianion in synthetic studies during the several preceding years. I quickly recognized that ring–metal bonding comparable to that in ferrocene would require an eight-lobed atomic orbital that I realized had to be one of the f orbitals (Figure 26).

The f orbitals immediately suggest the actinide and lanthanide elements. I mentioned this suggestion to Ullie Müller-Westerhoff°, who had been working with me for almost a year in continuing our fruitless attempt to prepare octalene. The hydrocarbon was later prepared in an elegant synthesis by Professor Emanuel Vogel at the University of Cologne.[334] Ullie had gotten his Ph.D. in Germany and had a long interest in organometallic chemistry. Thus, after all those months in unproductive organic synthesis, he was ready for a change and in one of his first experiments only a few days later treated cyclooctatetraene dianion

---

° *See* Appendix A for brief biographical information about each person designated with the ° symbol.

with uranium tetrachloride and obtained beautiful green crystals of the compound we then named uranocene. (eq 24).[335]

$$C_8H_8 \xrightarrow[\text{THF}]{\text{K}} C_8H_8^= \xrightarrow{\text{UCl}_4} (C_8H_8)_2U$$

"Uranocene"     (24)

We recognized that organometallic compounds of uranium(IV) were likely to be air-sensitive, and the synthesis was successful only because Ullie scrupulously degassed all solvents and apparatus.

Our success provided that joy of discovery that gives such unique delight to the life of a scientist. I recall having occasion to mention this discovery a few months later to the late Professor R. B. Woodward, who paid me the supreme compliment of remarking "I wish I had thought of that."

*Discussing chemistry with Professor Emanuel Vogel (right) in Cologne in 1991. Professor Vogel discovered one of the first stereospecific electrocyclic ring openings while a student with Criegee. He is best known for his syntheses and studies of the aromatic character of a variety of bridged cyclodecapentaene compounds, including his successful synthesis of "octalene", a compound we tried to prepare some years ago.*

## A Lifetime of Synergy with Theory and Experiment

Ken Raymond° had just been appointed, beginning July 1968, as an Assistant Professor at Berkeley. Ken is an inorganic chemist and an expert X-ray crystallographer. He recalls my telling him about our new uranocene synthesis over lunch at the Faculty Club one day that fall and asking about the feasibility of doing a structure determination. I was aware that uranium would dominate the X-ray diffraction and that the electron density about the much lighter carbon might be difficult to detect. Ken thought it might be feasible; however, at Berkeley he planned to downplay his crystal structure work and to emphasize other research interests. I thereupon took from my pocket a vial with some beautiful uranocene crystals, and he was hooked. He determined

*In 1978, we held a party to celebrate the tenth anniversary of the discovery of uranocene. Ken Raymond (left), who determined the crystal structure with Alan Zalkin, had these T-shirts made, which we wore at the party. My wife, Sue, had ordered a cake coated with green icing and the word "uranocene" written on it ten times. The bakery woman took the order while looking progressively dubious and finally asked, "That's not a dirty word, is it?"*

the structure in collaboration with Allen Zalkin, a research chemist and X-ray structure expert at the Lawrence Berkeley Laboratory (LBL), and proved the $D_{8h}$ structure with two planar octagonal rings having uranium at the center (Figure 27).

It is amusing that they had trouble publishing their result. A referee objected that the contribution of carbon to the diffraction compared to the much heavier uranium would be too small to determine, exactly the fear that had prompted me to consult Ken in the first place. Only when Ken pointed out to the referee that the symmetry of uranocene was such that the uranium could not contribute to many of the reflections was the communication published.[336] My vial of uranocene had a further effect on Raymond's research. He has continued with further studies in coordination compounds of actinides and, for example, in developing

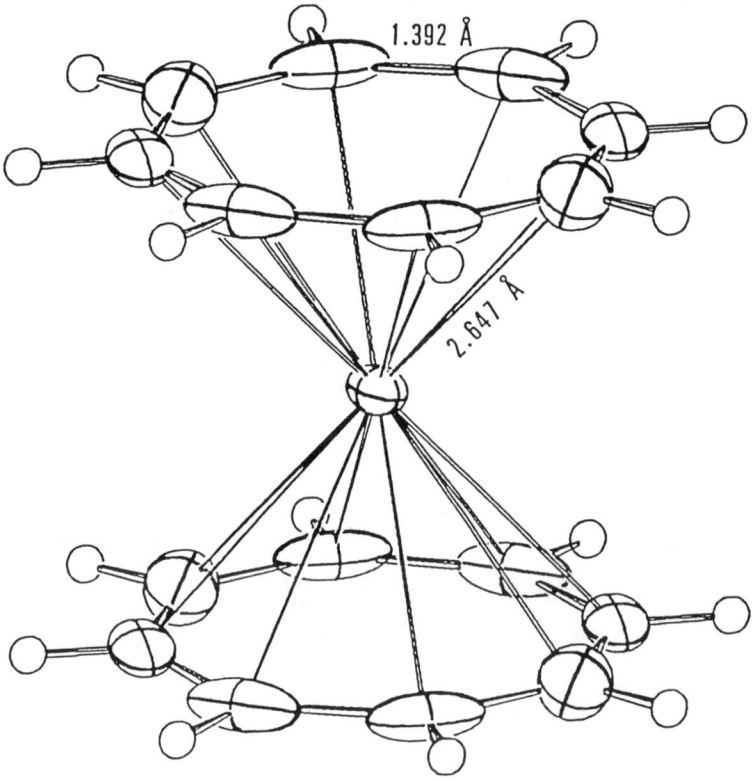

*Figure 27. X-ray crystal structure of uranocene. (Reproduced from reference 336. Copyright 1969 American Chemical Society.)*

# A Lifetime of Synergy with Theory and Experiment 187

sequestering agents for plutonium that could be used for treating plutonium poisoning in humans.

Until the discovery of uranocene, organoactinide chemistry was virtually a virgin field. Gilman had directed research on organouranium chemistry during World War II as part of the Manhattan project, and some of this work was later published.[337,338] Cyclopentadienyl derivatives of actinide elements had been

*Speakers at a Natick Conference in October, 1969, on frontiers in chemistry. Back row (from left): Ernest Eliel, Alan W. Johnson (Manchester University, now deceased), Duilio Arigoni (ETH, Zurich), A. Streitwieser, and Louis Long, Jr. (conference organizer). Front row (from left): Roald Hoffmann, Max Muxfeld (Cornell University, now deceased), and Siegfried Hunig (University of Wurzburg). I talked about uranocene.*

prepared in 1956.[339] I learned later of a theory paper by R. D. Fischer[340] that discusses the possibility of the uranocene structure but without recognition of the $e_{2u}$–$f_{\pm 2}$ ring–metal bonding scheme.

After the publication of the synthesis of uranocene the general field of organoactinide and -lanthanide chemistry expanded rapidly. Several laboratories pursued the cyclopentadienyl ligand and greatly expanded this chemistry.[341] In Berkeley, there was enough to do in following up the chemistry of the cyclooctatetraene ligand. Ullie Müller-Westerhoff finished his postdoctoral year and left Berkeley shortly after his synthesis. He had committed himself to a research position at IBM and later joined the faculty at the University of Connecticut. His place was taken quickly by other postdoctoral co-workers and graduate students, and our entry into this field developed in several directions. George Sonnichsen confirmed the synthesis and pioneered the early chemical studies.

This early group included Frank Mares, for whom this was also a brand new research area. He had been a research chemist at the Institute for Process Fundamentals, Prague, Czechoslovakia, with expertise in organosilicon chemistry. He obtained a leave of absence for 1966–1967 to do postdoctoral work with me and obtained some important carbon acidity results, particularly with silicon and fluorine compounds. Shortly after his return to Prague, the Czechoslovakian revolt occurred and the resulting Russian invasion. Frank was able to leave the country with his family. They were able to stay for some time with Günther Häfelinger, who had also been a postdoctorate in my group and was now teaching in Tübingen. Frank Mares eventually made his way back to Berkeley where I was able to provide postdoctoral support until he obtained a permanent position. During this second period he did his important work in organoactinide chemistry, made the first lanthanide derivatives of cyclooctatetraene, and developed new techniques for this chemistry. He was part of my group at that time that made a movie of these techniques to help train new students. This group also included Kieth Hodgson°, who learned X-ray crystallography from Ken Raymond and solved a number of crystal structures. In effect, although formally an organic chemist, he worked jointly with both of us.

The initial synthesis was developed into a full paper[342] and later provided as an *Inorganic Synthesis* preparation.[343] Norio Yoshida spent one year in my laboratory on a leave of absence from his Japanese company, Teijin, Ltd., and extended the synthesis to the thorium analog.[344] I have found it convenient to refer to these compounds as [8]annulene metallocenes; thus, uranocene is bis-([8]annulene)uranium(IV). The specification that the uranium is in a formal oxidation state of +4 then implies, because the compound is neutral, that the [8]annulene is present as the dianion. Other formal names have been proposed, but this one seems to me to be the most general and straightforward. Shortly thereafter, the Karraker group at Savannah River prepared the neptunium and plutonium compounds[345], and we prepared bis-([8]annulene)protactinium.[346,347]

We also made a number of substituted uranocenes by starting with cyclooctatetraenes substituted with alkyl, aryl, alkoxy, amino, and carboxylate functions.[348–352] These derivatives allowed systematic studies of physical and chemical properties.[353] The objective of exploring a range of chemistry was to seek materials with useful physical properties or reactions that could lead to useful reagents and catalysts. We soon learned that the reaction chemistry of uranocene derivatives is not rich. Ferrocenes are useful compounds because they show typical aromatic substitution reactions and the compounds are stable to air and moisture. Uranocene does not undergo electrophilic aromatic substitution. Mild reagents have no effect, and strong electrophilic reagents decompose the molecule. A few interesting reactions were discovered,[354,355] but the reaction chemistry is disappointingly meager. Some of the physical properties, notably NMR, however, have special interest.

One of the most interesting and useful aspects of uranocene chemistry is its paramagnetism. Two 5f electrons are left on uranium(IV), and these are unpaired. I was unfamiliar with this aspect of inorganic chemistry and, in 1970, began a collaboration with Norman Edelstein, a research chemist at the Lawrence Berkeley Laboratory and an expert in magnetic phenomena and spectroscopy. A new graduate student at that time, Dennis Morrell, studied the magnetic susceptibility of some uranocene derivatives.[356]

*Sue with Jenny Green and Wilhelm Maier at lunch in the Napa Valley, California, in 1981. Jenny Green is a physical chemist at Oxford who determined the first photoelectron spectra of uranocenes. She and her husband, Malcolm Green, were our hosts during our two-week stay in Oxford in 1980. Wilhelm Maier was an Assistant Professor at Berkeley at that time; we collaborated in getting an NSF grant for a VAX computer that we shared effectively. He is now at the Max Planck Institute in Mülheim, Germany.*

The next step was to study the NMR spectrum. Here, we were helped by Gerd LaMar°, an expert in paramagnetic NMR then at the Shell Development Company, Emeryville, California. Not long thereafter Shell moved their laboratory to Houston and reduced their staff. Gerd found himself temporarily out of a job. He had obtained a faculty position at the University of California at Davis, but that would not start for six months. Fortunately, I was able to hire him as a postdoctorate for the intervening period. The most valuable NMR spectrum at that time was that of the uranocene derivative prepared from 1,3,5,7-tetramethylcycloocta-tetraene.[348] In this compound, we could compare the chemical shifts of ring and methyl protons. I asked Gerd LaMar many questions trying to understand something about paramagnetic NMR. In the process of trying to answer these questions he suddenly realized that he had made a wrong assumption about spin interactions in these compounds that led to the wrong direction of the chemical shift. With the assumption of spin polarization in charge transfer bonding from ligand to metal, the correct chemical shift directions were derived.[348] These spin polarization changes are summarized in Figure 28. I also learned enough to use this NMR as a tool for studying various types of effects in the uranocene system.

The magnetic moment associated with the unpaired electrons on uranium affects the magnetic resonance of all of the nuclei. For example, the $^1$H NMR chemical shift of the ring protons of uranocene occurs at $\delta = -36.6$ ppm at 30 °C, well upfield from the normal range of $^1$H NMR spectra. Part of this shift comes from a through-space effect, in which the uranium is behaving as an internal paramagnetic shift reagent, the so-called pseudo-contact or dipolar shift, and part comes from a spin-polarization of electrons in bonds, the so-called contact shift.[357] The contact shifts come from the spin polarizations symbolized in Figure 28. This shift has a different sign for ring protons and for α-protons. The chemical shift changes with temperature, and the lines are broad-

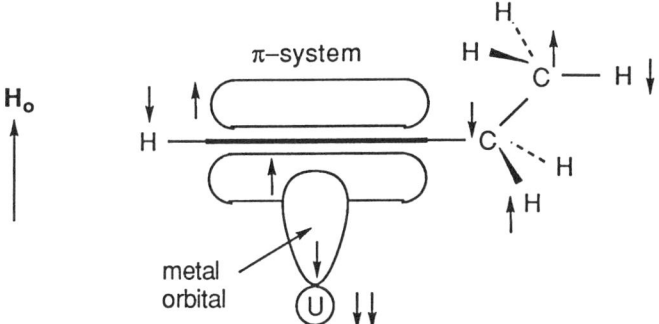

*Figure 28. Spin polarizations in a substituted uranocene. Arrows shown refer to magnetic moments. The two 5f electrons in U(IV) have the same spin (Hund's rule), and their spin magnetic moments are opposed to the applied field in the ground state because of the large orbital magnetic moment that gives a total moment aligned with the field. Electron density donated from the ligand π-system to any empty orbital on U is spin polarized so that the donated spin density close to the 5f electrons has the same spin. This leaves an excess of the opposite spin in the π-system. Further spin polarization in the δ system leaves an excess of the opposite spin in the ring-H bond close to H corresponding to an upfield contact shift. (Reproduced from reference 357. Copyright 1980 American Chemical Society.)*

ened somewhat by the paramagnetic effect but are still sharp enough for most use. The broadening does, however, obscure the spin–spin splitting in most substituted uranocenes.

Structure has a large effect on the chemical shifts of ring protons. For example, in 1,1′-dimethyluranocene the ring protons now have $\delta$ = –31.70, –33.67, –36.10, and –40.39; this range is equivalent to almost the whole normal range of proton chemical shifts! The methyl protons have $\delta$ = –7.2 ppm, showing a large downfield contact shift that counteracts the upfield dipolar shift. The assignment and understanding of these chemical shift patterns was a significant research effort, and the large effect of structure on chemical shifts has important applications to conformational and dynamic NMR studies. For example, in 1,1′-dineopentyluranocene, the *tert*-butyl methyl groups are sufficiently far from the ring that contact shift effects are negligible. With some simplifying assumptions, the dipolar shift can be calculated from the structure, and the observed $\delta$ = 3.9 ppm agrees with an *exo-tert*-butyl group conformation. Similarly, the variable-temperature NMR of 1,1′-diethyluranocene indicates that at least two conformations are present, but an exo-conformation is prominent. Many of these conformational studies were carried out by Stuart Berryhill and Wayne Luke and summarized by Wayne in a thorough review for an American Chemical Society symposium.[357]

The large chemical shift changes for different protons allow the facile determination of dynamic effects even with relatively low barriers. An interesting and typical example is the ring rotation in 1,1′,4,4′-tetra-*tert*-butyluranocene.[358] One ring has three different types of ring protons, and at 30 °C these correspond to three equal resonances at $\delta$ = –25.23, –39.66, and –42.23. At about –70 °C, however, these peaks broaden in typical NMR coalescence fashion, and at still lower temperature they reemerge as six peaks. In the same manner, the sharp singlet at 30 °C for the *tert*-butyl group at $\delta$ = –10.25 splits into a doublet at low temperature. These changes are shown in Figure 29. Although it was not possible to assign the ring protons, the splitting into six peaks at low temperature excludes the symmetrical conformation shown as a Newman-type diagram, **31a**. In this structure we are looking down the ring–uranium axis, with the dotted *tert*-butyl groups

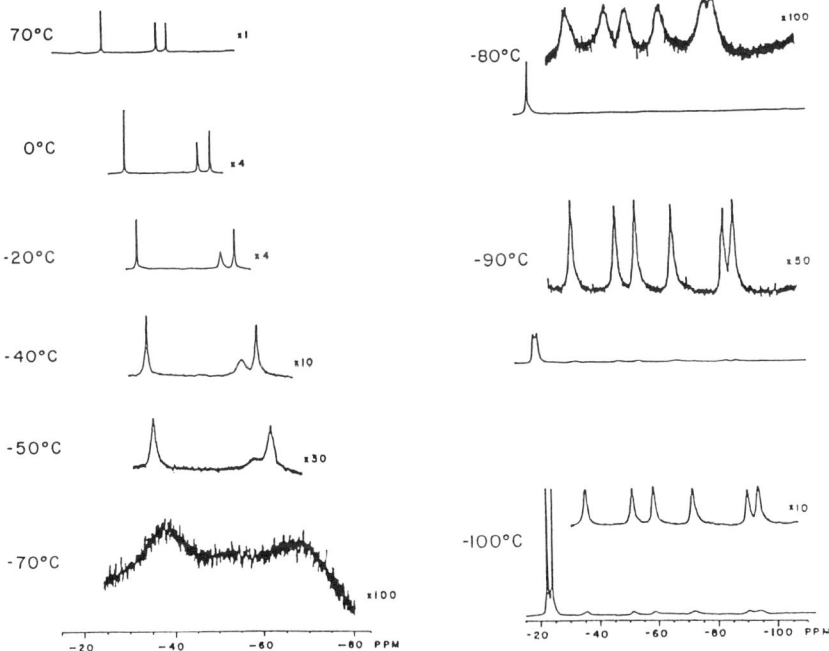

*Figure 29. $^1$H NMR spectra of 1,1´,4,4´-tetra-tert-butyluranocene as a function of temperature. The spectra at the left show only the ring protons. The spectra at low temperature show the tert-butyl group as well. (Reproduced from reference 358. Copyright 1981 American Chemical Society.)*

belonging to the lower ring. Instead, the stable conformations are similar to **31b** and **31c**. At higher temperatures we see only the averaged spectrum, but at low temperatures the six ring hydrogens are magnetically different and there are two types of *tert*-butyl groups. From the coalescence temperatures the barrier to ring rotation could be estimated as $\Delta G^{\ddagger}$ = 8.3 kcal/mol. A similar study of the diamagnetic 1,1′,3,3′-tetra-*tert*-butylferrocene gave much smaller splittings, but the barrier, $\Delta G^{\ddagger}$ = 13.1 kcal/mol, was large enough to give measurable coalescence. The higher barrier for the ferrocene compound reflects the shorter ring–ring distance in ferrocenes.

Another example relates to the barrier to rotation of a phenyl group in aryluranocenes. The ortho-phenyl hydrogens in 1,1′-diphenyluranocene show an NMR resonance at $\delta$ = –13.66 ppm at 30 °C. At low temperatures this peak broadens, but we were not able to get to low enough temperatures to see coalescence. The phenyl substituent is twisted with respect to the eight-membered uranocene ring, and the exo- and endo-ortho hydrogens, **32**, have different chemical shifts. For example, in an *o*-tolyluranocene, in which the methyl group remains in the exo-position, the endo-ortho ring hydrogen is close to the uranium center and has $\delta$ = –26.2 ppm at 30 °C. If we estimate that the endo-proton in **32** has the same value, then the exo-proton must have $\delta$ = –1.2 ppm in order for the average value to be that found experimentally. From such results we could model the line shapes as a function of temperature to derive an enthalpy of activation for rotation of the phenyl group, $\Delta H^{\ddagger}$ = 4.4 ± 0.3 kcal/mol.[359] Measuring a dynamic process with such a low barrier as in this example is unusual but is made possible by the paramagnetism of uranocene that puts these protons in such different magnetic environments. For comparison, the analogous rotation of a phenyl group in biphenyl has a barrier of 1–2 kcal/mol.[360] The higher barrier in uranocene undoubtedly results from the wider C–C–C bond angle in the [8]annulene ring compared to benzene. These examples are indicative of the usefulness of dynamic NMR studies of uranocenes. Because of its particular paramagnetic properties and well-defined structure, I think that this general structure could be useful for defining a number of conformational and related effects in organic compounds.

**32**

My association with Norm Edelstein and the Nuclear Division of the Lawrence Berkeley Laboratory led to further use of their facilities. Chemical studies in this division included the chemical and physical properties of transuranium actinides that evolved from the pioneering studies of Glenn Seaborg.° With the special facilities available at LBL to work with highly radioactive materials, one of my students, Dave Starks, developed a new preparation of the plutonium analog of uranocene[361] and prepared a small amount of the corresponding protactinium compound.[346,347] Work with such elements is rather special and is done in individual "hot boxes" for the different elements. Research by one person is done under the watchful eyes of a separate monitor. Careful attention is paid to exposure levels of radiation. This type of research affords the opportunity for unique experimental experience, and some students derive a special joy from it.

Although this research was supported at that time by the NSF, our work was incorporated into LBL reports. In 1975, chemical research at LBL underwent a reorganization. Much of the actinide chemistry work became part of a new division, the Materials and Molecular Research Division (MMRD) (recently reorganized as the Chemical Sciences Research Division, CSRD), and both Ken Raymond and I became formally a part of this new Actinide Chemistry group. In 1976, when Dick Andersen° joined the faculty, he also became part of this group. We held an interdisciplinary seminar on a biweekly basis at LBL to share our research results. Work with uranium itself requires only normal precautions; the uranium currently available is "depleted", with the $^{235}$U isotope removed, and has consequently only slight radioactivity. Most of our uranium chemistry was done on campus, but the LBL facilities were essential for the research with transuranium elements that had been a significant part of my f-element organometallic research.[362,363] This research led most recently to NMR studies of cyclooctatetraene derivatives of neptunium and plutonium in the III and IV oxidation states, the first NMR spectra of transuranium organometallic compounds.[363]

I am fond of citing this wide range of experimental and theoretical work in my research group to those who fear that modern science is becoming too narrow and specialized. Unfortunately, my research support from DOE was terminated in 1990, appar-

*My inorganic colleague Richard A. Andersen. Dick's organometallic chemistry includes research with organouranium compounds. Dick and I attended a NATO conference on f-element organometallic chemistry in Maratea, Italy, September, 1984. We left together early because of teaching commitments at Berkeley and spent the night en route in Rome. We had a few hours for sightseeing and visited the* Boca Verita *(Mouth of Truth). Dick tested the legend according to which if one were unfaithful one's hand would be chopped off. Dick was able to retrieve his hand unscathed.*

ently because we were not sufficiently active in transuranium actinide chemistry (despite the discoveries mentioned already). Meanwhile, the rules and regulations (and legal liabilities!) for radioactive research became progressively more onorous so that I recently stopped all such work on campus and cleaned out all of our radioactive samples. This meant disposing of a large number of NMR samples of various substituted uranocenes, a valuable and useful inventory, even though the total radioactivity involved was less than that present in one gas light mantle (which contains thorium oxide) available by anyone without restriction from most hardware stores.

The preparation of thorocene was followed up by a Miller postdoctoral fellow, Carole LeVanda, who prepared several substituted thorocenes.[364] Thorium(IV) has no 5f-electrons and has a closed-shell electron structure. Thorocenes are therefore diamagnetic, and their NMR spectra serve as valuable diamagnetic reference systems for the analogous uranocenes. She also prepared the half-sandwich compound 33 and established the equilibrium in eq 25.[365]

$$\text{Th(COT)}_2 + \text{ThCl}_4 \rightleftharpoons 2 \;\; \underset{33}{\text{Th(COT)Cl}_2} \quad (25)$$

Two half-sandwich structures were established by X-ray diffraction.[366] Uranocene does not undergo such an equilibrium. On reaction with $UCl_4$ thorocene gives uranocene, but not rapidly. Uranocene appears to be the most stable of these sandwich structures, and thorocene appears to be distinctly more ionic.

Reaction of thorium tetrachloride with one mole of cyclooctatetraene dianion gives the half-sandwich directly. This method does not work with uranium; one gets just a 50% yield of uranocene! Several methods were subsequently developed for preparing uranium half-sandwich compounds, and their structures were determined.[367] We had hoped to be able to replace the chlorines with alkyl groups to determine their potential as catalysts, but the program was terminated before this goal was accomplished. One of the most interesting products, however, was [8]annuleneuranium(IV) bis-acetylacetonate, $(C_8H_8)U(acac)_2$, whose structure is given in Figure 30.

Much of this work in the uranocene half-sandwiches was done by my last graduate student in organouranium chemistry, Tom Boussie. Tom's versatility is demonstrated by the extensive organic synthesis research he did in developing new routes to substituted cyclooctatetraenes[368] even though he is formally an inorganic chemist.

One of our final results was the preparation of the first bridged uranocene, [1, 2-bis-([8]annulenyldimethylsilyl) ethane]-uranium, **34**.[369] This research was done by two visiting professors, Dr. Hsu-Kun Wang of Nankai University, Tianjin, People's Republic of China (now at the University of Utah), and Dr. Maria Teresa Barros of the New University of Lisbon, Portugal. Both

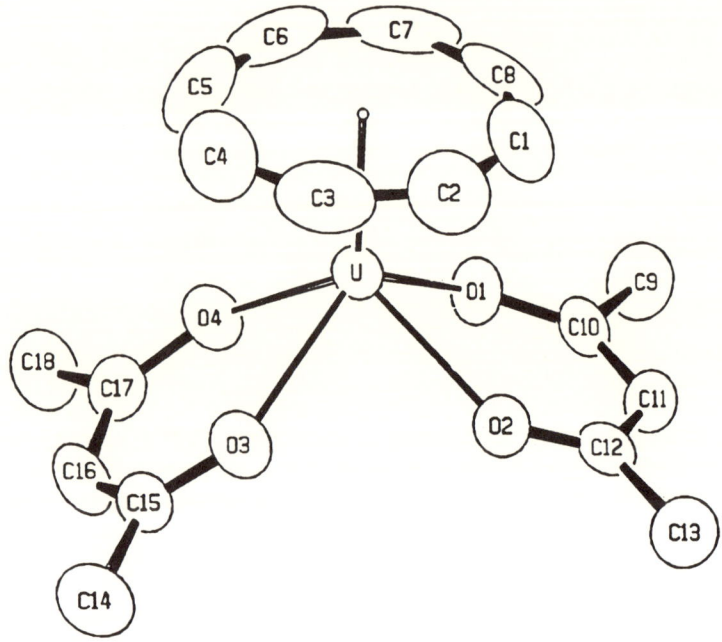

*Figure 30. X-ray crystal structure of $(C_8H_8)U(acac)_2$. (Reproduced from reference 367. Copyright 1990 American Chemical Society.)*

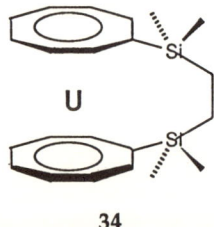

34

women have spent several periods in my laboratory. Hsu-Kun spent the 1979–1980 year with me and another year with my late colleague, Earl Muetterties. She worked on many projects in uranocene chemistry. In the summer of 1981, I spent two weeks in China, giving a short course on MO theory in organic chemistry at Beijing University. Hsu-Kun served as my interpreter. Each 1½-hour lecture lasted for close to three hours because all of my English had to be translated into Chinese. In such presentations it

# A Lifetime of Synergy with Theory and Experiment

At the Second International Conference on Lanthanides and Actinides, Lisbon, April, 1987. From left: Notker Rösch, Andrew Streitwieser, Patricia Watson (a DuPont chemist who discovered one of the first "hydrocarbon activation" reactions of organolanthanides), my LBL colleague and collaborator Norman Edelstein, and Hsu-Kun Wang. Professor Wang spent a year in my laboratory, 1979–1980, and was my translator when I gave a short course in molecular orbital theory in Beijing, China in 1981.

was customary for the speaker to talk for a whole paragraph or more and then wait while the whole passage was translated, but Hsu-Kun was so well acquainted with my speech patterns that she was able to translate sentence by sentence and even phrase by phrase. I learned later that many in the audience greatly appreciated her facility because they knew some English and her quick translation provided instant reinforcement. Teresa Barros has visited Berkeley several times in the company of her husband, Chris Maycock, a synthetic organic chemist who is also a professor in Lisbon. Chris was a former postdoctoral co-worker of Henry Rapoport and returns periodically in their continuing collaboration. They were my hosts in a delightful tour of Portugal following an f-element meeting in Lisbon in 1987.

## Organolanthanide Chemistry

Shortly after the preparation of uranocene we studied the analogous compounds of the lanthanides. In their compounds, the lanthanide elements exist mostly in the +3 oxidation state. We prepared a number of cyclooctatetraene derivatives by the general type of reaction shown in eq 26 for $Ce^{3+}$.[370,371]

$$C_8H_8^{2-} \cdot 2K^+ + CeCl_3 \rightarrow (C_8H_8)_2Ce^- \cdot K^+ \qquad (26)$$

The products are sandwich structures with a net single negative charge and require a gegenion such as potassium cation. With appropriate stoichiometry, half-sandwich compounds form readily and crystallize as dimers, **35**.[371] X-ray structures were obtained[372], and several of the lanthanide analogs were compared by their physical properties and spectra. These compounds are essentially ionic aggregates of the trivalent cations and the cyclooctatetraene dianions.

**35**

Some lanthanides, particularly samarium, europium, and ytterbium, have readily accessible +2 oxidation states. One of Steve Kinsley's achievements as a graduate student was to show how a related sandwich structure, $(C_8H_8)_2YbK_2$, could be prepared from $Yb^{2+}$.[373] $Yb^{2+}$ has a filled 4f shell and is diamagnetic; the sandwich structure has a normal NMR spectrum. However, the corresponding compound of $Yb^{3+}$, $(C_8H_8)_2YbK$, is paramagnetic, and the proton NMR is so broad as to be invisible. A mixture of the two compounds shows no NMR signal at room temperature because of rapid electron transfer on the NMR time scale. At low temperatures the rate of electron transfer is slow enough that the ring protons of the $Yb^{2+}$ compound are again visible. With suitable derivatives it was possible by line-shape analysis to obtain rates of the electron transfer process.[374] The rates are fast but are less than diffusion controlled. We showed that the exchange process involved was that of electron transfer rather than ring exchange by a simple scheme. Calcium has the same ionic radius as $Yb^{2+}$. It gives a comparable cyclooctatetraene derivative, $(C_8H_8)_2CaK_2$, which shows no exchange with $(C_8H_8)_2YbK$ on the NMR time scale because Ca has no accessible +3 valence state. This was, to my knowledge, the first preparation of a calcium derivative of cyclooctatetraene.

As I mentioned already, we had studied the NMR spectra of the neptunium and plutonium compounds of cyclooctatetraene with the metals in the +4 and +3 oxidation states; these metals readily form compounds in a variety of oxidation states. The +3 oxidation state of uranium is also common, and we and others tried for some time to prepare the corresponding cyclooctatetraene compound. Dave Eisenberg, another inorganic student working with me, finally succeeded, not only in preparing $(C_8H_8)_2U^- K^+$ by reduction of uranocene and characterizing it by NMR,[363] but in obtaining crystals of a derivative for X-ray analysis.[375] The compound is exceedingly air-sensitive and behaves as a more ionic compound than uranocene. The NMR spectra of mixtures show rapid electron transfer between all of the cyclooctatetraene derivatives of the +3 and +4 actinides.[363]

## Cerocene

The simple theoretical idea that led to the synthesis of uranocene has opened a wide area of chemistry, but this result by itself does not prove that the generating idea was correct, namely that f orbitals are important in ring–metal bonding of cyclooctatetraene derivatives of the actinides. It is actually impossible to "prove" such a theory because it is based on qualitative concepts not grounded in physical observables. The best one can do is to compare whether a range of chemistry is consistent with the concept. The same point was made in a slightly different way in the previous discussion of ab initio calculations. It is an important aspect of the philosophy of science that I hope I instill in my students. Its essential feature is that chemistry, the properties and reactivities of substances, is a highly complex science that can only be understood by dissection into parts, each of which can be analyzed, used for predictions, and understood in the sense of a model. These parts, effects, and models can never be "proved" as entities apart from the whole but do provide an essential systematic framework that can help mere humans to assimilate and manipulate a body of experimental fact.

A number of types of results indicate that the original simple theory that started this chemistry is substantially correct. For example, Mössbauer spectroscopy of neptunocene shows greater charge transfer from the cyclooctatetraene ligands to the central metal than for any other $Np^{4+}$ compound studied.[345] Photoelectron spectra of thorocene and uranocene indicate involvement of 5f orbitals in ring–metal bonding but also show involvement of 6d orbitals to at least as great a degree in a manner similar to that in the d-transition metal metallocenes.[376–379] The bulk of the physical and chemical evidence seems to agree that there is significant ring–metal covalency in these cyclooctatetraene derivatives of the actinides and that this covalency does involve f orbitals as in my original thought, but that d orbitals are also important.[380]

An amusing and educational development should be added. In 1976, an Italian group reported the preparation of a neutral sandwich compound of cerium(IV), bis-cyclooctatetraene cerium(IV).[381] They reported complete experimental details and adequate characterization of the compound, but I must confess that I found it difficult to understand this result. Lanthanide compounds are generally considered to be highly ionic. Thus, as indicated already, the bonding interaction between the dianion ligands and the central +3 lanthanide cation in the [8]annulene sandwich compounds is thought to be primarily electrostatic with but little covalency. Cerium(IV) is a powerful oxidizing agent, sufficiently so in aqueous solution to oxidize chloride ion to chlorine, and it did not seem likely to be able to exist in an ionic assembly with the strong reducing agent cyclooctatetraene dianion. My further involvement with this matter, however, resulted from a research collaboration in Germany.

In 1975, I was nominated for a Humboldt Award for Senior U.S. Scientists by Professor Ivar Ugi of the Technical University, Munich. I received the award and was able thereby to spend the first half of 1976 in Munich as Professor Ugi's guest. The Humboldt awards and related fellowships are administered by the Alexander von Humboldt Foundation of Bonn, Germany. The eponymous Alexander von Humboldt (1769–1859) was a world

# A Lifetime of Synergy with Theory and Experiment

*I love Munich beer as this picture with "eine Mass" (a one-liter container) indicates. A beer garden at the Chinese tower in the English Gardens is popular with foreigners and Müncheners.*

famous explorer and scientist. The Foundation was established by friends after his death and was reestablished as a private foundation with financing by the Federal Republic of Germany in 1953. Fellowship support for foreign scholars in the sciences and humanities was provided for research in Germany, and in recent years German scholars have been supported for work in other countries. More than 14,000 postdoctoral researchers have been supported by such fellowships during the past four decades.

*Professor and Mrs. Ivar Ugi and their dog in 1991. Professor Ugi nominated me for my Humboldt award and hosted our six-month stay in Munich in 1976.*

On June 5, 1972, Chancellor Willy Brandt announced the establishment of the U.S. Senior Scientist research awards (Sonderpreis) to mark the 25th anniversary of the speech by General George C. Marshall at Harvard in which he proposed the postwar aid to Europe now known as the Marshall Plan. The implementation of this plan by Congress the following year was an essential factor in the rebuilding of the Federal Republic of Germany after the devastation of World War II. These competitive awards were made to recognized U.S. scientists to work at German universities and research institutes for periods up to one year. The new awards were administered by the Humboldt Foundation in addition to their normal fellowship program. The first such awards were made in 1972. Thus, I was one of the early recipients of this award. Since then, more than 1400 professors and industrial scientists have been recipients of this award and have carried out research projects in Germany. The University of California, Berkeley, has the largest number of award winners, more than 60. In recent years these awards have been extended to other countries.

*Johnny and Ullie Gasteiger at a dinner in 1995. They helped us to get settled in Munich for my 1976 Sabbatical leave. He took Ullie as his brand-new bride to Berkeley in 1971 to spend a postdoctoral year with me doing both ab initio calculations and experimental research. Johnnie is a major figure in computational chemistry and is now director of a computational chemistry laboratory at the University of Erlangen–Nürnberg.*

Thus, Sue and I spent six months in Germany in 1976, the longest period that I have been away from Berkeley since my arrival there in 1952. In preparation, both Sue and I took German lessons the preceding semester. I had taken German in high school and college, but my knowledge of the language was rusty and the lessons were a great help. I wish my command of the language was fluent, but my knowledge is adequate for many purposes. Our visit in 1976 was a wholly fascinating and delightful experience for both of us. We learned about and enjoyed many local customs, and I learned a great deal about the German educational system, as well, of course, as meeting many chemists and benefiting from mutual research discussions.

I mentioned earlier my meeting of Professor Hückel in Marburg. I also met Dr. Heinrich Pfeiffer, Secretary-General of the

*Sue and I bought sheepskin coats in preparation for my Sabbatical in Germany in 1976. We arrived on New Year's Day, and the coats were invaluable for a cold European winter much different from winters in Berkeley. On this occasion we are enjoying that great Bavarian beer.*

Humboldt Foundation from 1956 until his retirement in 1994, to begin what has been a long and friendly relationship ever since. My association during our Munich stay was primarily with the research group of Professor Ivar Ugi, but space was tight in 1976. The chemistry institutes were located in old buildings in the heart of Munich; not long after, these institutes moved to a large new building north of Munich near the town of Garching. Office space in the old complex was provided for me through the hospitality of Professor Ludwig Hofacker at his Lehrstuhl für Theoretische Chemie. As a result, I met many of the theoretical chemists and discussed chemistry with them.

One of these theoreticians is Professor Notker Rösch, who had previously spent a postdoctoral period in the United States during 1972–1974 with Keith Johnson at MIT and then at Cornell with Roald Hoffmann and Thor Rhodin. Roald Hoffmann's group was already doing theoretical studies of organometallic compounds at that time, but Notker contributed to the first paper of

# A Lifetime of Synergy with Theory and Experiment

A seminar at a German university is generally followed by a Nachsitzung, dinner at a local restaurant. This Nachsitzung followed a seminar I gave at the University of Munich in 1991. From left: Professors Knorr, Huisgen, and Gompper, my wife Sue, and my former postdoctoral co-worker, Johnnie Gasteiger.

*Heinrich Pfeiffer (left), Secretary-General of the Humboldt Foundation, with Sue and me during the 25th anniversary celebration of the reestablishment of the Humboldt Foundation, December, 1978.*

the group in this area, on tris(ethylene)nickel. He told me of his work with Keith Johnson applying the "Xα scattered wave" method to ferrocene, and I immediately wondered about the possible applications of this approach to uranocene. Notker was also interested in my uranocene story and readily accepted my suggestion that we try it out. This beginning led to a collaboration that resulted in the successful application for a joint NATO grant. During several trips between Munich and Berkeley, we carried out Xα MO calculations of uranocene and thorocene.[382]

The work required that Notker modify his program to accept f orbitals. This he did, but these first calculations were done nonrelativistically. With heavy elements, such as uranium, electrons close to the nucleus have an effective velocity that approaches the speed of light and require a relativistic correction. Notker was subsequently able to modify his program to incorporate relativistic effects, and these corrections were included in the second series of calculations.[383] Notker later was able to rationalize why the nonrelativistic results gave such good agreement.[384] The

# A Lifetime of Synergy with Theory and Experiment

*Professor Paul D. Bartlett is on my right at a meeting of the Humboldt Foundation when we were both recipients of Senior Scientist Awards in 1976. Paul is one of the major figures in modern physical organic chemistry. He served on the faculty of Harvard University from 1934 to 1974. After his retirement from Harvard he started a second career as Welch Professor at Texas Christian University. In the summer of 1977, we were again in Germany together, and we and our wives had a delightful drive to Innsbruck, Austria..*

good agreement between calculated and experimental ionization potentials supports the model that significant bonding involves the f orbitals exactly as I had originally proposed. These results now form a substantial part of the overall picture of the electronic structure of these actinide sandwich compounds. More recent calculations[380] have essentially confirmed these results; the only difference concerns the relative importance of bonding with d and f orbitals. The interaction of d orbitals as in the d-transition metallocenes is clearly at least as important as that of the f orbitals.

With these promising results on uranocene and thorocene, I suggested to Notker that he extend the calculations to the related cerium(IV) compound to which I gave the name of cerocene. The

*Professor Ludwig Hofacker at a luncheon meeting of his Lehrstuhl für Theoretische Chemie at the Technical University, Munich, in 1991. Professor Hofacker generously provided office space for me during my frequent visits to the university.*

results indicated that the compound is completely normal! That is, the calculations indicated a sufficient ring–metal covalency in this compound to prevent the complete electron transfer of a net oxidation–reduction reaction. For example, part A in Figure 31 shows electron transfer from an occupied high energy orbital on encountering a lower energy empty orbital. An example is the electron transfer that occurs when a sodium atom approaches a

# A Lifetime of Synergy with Theory and Experiment

*In my temporary office in Professor Hofacker's Lehrstuhl für Theoretische Chemie in the new chemistry building of the Technical University, Munich, just outside the town of Garching north of Munich, in 1977.*

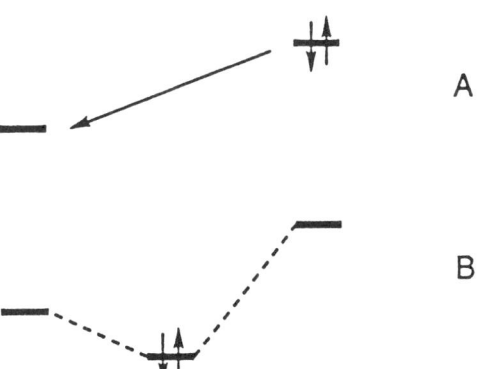

*Figure 31. A. An occupied orbital of higher energy in one fragment will donate charge in a simple oxidation–reduction reaction to an empty lower orbital of another fragment with which there is no effective overlap. B. Overlap produces a bonding orbital between the two fragments equivalent to only partial charge transfer from donor to acceptor.*

*With Frau Ingrid Braun, Professor Hofacker's secretary for many years, in 1985. Frau Braun did many kindnesses for me during my frequent stays in Munich. Although we frequently talk together in German, her English is excellent. I recall one occasion when Professor Hofacker, as editor of* Chemical Physics, *was writing to one of his German colleagues. Frau Braun asked him whether she should use the* Sie *or the* du *form of address. Hofacker pondered awhile and replied, "Write him in English."*

fluorine atom. On the other hand, with sufficient overlap the two orbitals form a bonding MO as in B in Figure 31. The electrons in this bonding MO in effect correspond to incomplete charge transfer and to covalent bonding.

# A Lifetime of Synergy with Theory and Experiment

*Professor Notker Rösch in 1978. Notker is a theoretical chemist at the Technical University, Munich. We collaborated for several years in calculations that confirmed the significant role of bonding with f-orbitals in uranocene.*

After receiving Notker's surprising results of the calculations, I suggested to Steve Kinsley that he try to repeat the Italian preparation. He did so and in June 1983, confirmed their preparation in every detail. This route made the parent compound available, but because of its use of cyclooctatetraene as the solvent this method is not practical for the preparation of derivatives. Thus, we sought a suitable oxidation of the readily available Ce(III) compounds. Strong oxidizing agents gave cerocene but in low yield. It was the determination of redox potentials by cyclic voltammetry (an experiment suggested by Ken Raymond) that pointed out the problem: cerocene is almost 3 V weaker as an oxidizing agent than is aqueous $Ce^{4+}$; indeed, cerocene is more stable than the Ce(III) analog relative to the hydrogen electrode, and we had been using oxidizing agents that were far too strong! Shaking a solution of the Ce(III) compound with silver iodide sufficed to give cerocene (eq 27);[385] this method was used to prepare several derivatives.

$$[(C_8H_8)_2Ce]K + AgI \rightarrow (C_8H_8)_2Ce + KI\downarrow + AgI\downarrow \qquad (27)$$

We have found more recently that this method tends to be erratic; the method involves a heterogeneous reaction and depends on the nature and history of the solid. Overoxidation can occur. The same problem has been found with some other solid reagents. However, allyl bromide turns out to be an excellent reagent for this purpose. Undoubtedly, a one-electron oxidation is involved to give allyl radical that dimerizes to give biallyl (Scheme X). We have prepared substantial amounts of cerocene and derivatives by this route.[386]

$$[(C_8H_8)_2Ce]\ K\ +\ CH_2=CHCH_2Br$$
$$\downarrow$$
$$(C_8H_8)_2Ce\ +\ KBr\downarrow\ +\ [CH_2=CHCH_2\cdot]$$
$$\downarrow$$
$$CH_2=CHCH_2CH_2CH=CH_2$$

*Scheme X.*

We were able to provide a sample of cerocene to Professor Fragala of Catania, who measured its photoelectron spectrum. Good agreement was found with Rösch's computed ionization potentials, confirming the importance of covalent bonding in this compound.[385] One of our most recent results is the X-ray crystal structure determination of 1,1´-dimethylcerocene that shows its sandwich uranocene-like character (Figure 32).[375] Thus, a type of chemistry that I did not believe could exist a few years ago became an important part of my own research. Moreover, Professor Rösch has since been actively engaged in his own independent research program on lanthanide chemistry. The results of my

## A Lifetime of Synergy with Theory and Experiment

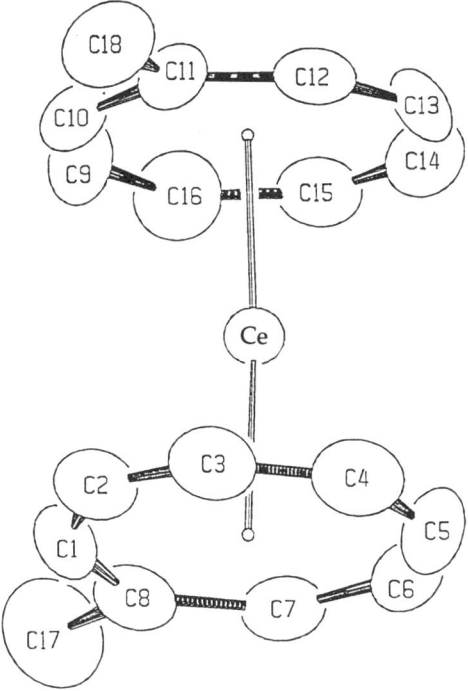

*Figure 32. X-ray crystal structure of 1,1´-dimethylcerocene. (Reproduced from reference 375. Copyright 1991 American Chemical Society.)*

collaboration with Notker Rösch provide an excellent example of the value of international scientific travel and of programs such as the NATO research grants and of the Humboldt Sonderpreis program.

Notker Rösch was also recently involved in one of those stories that indicate how the world revolves in small circles. Roald Hoffmann has become a poet in recent years.[387] To hone his poetry skills he spent some time at Djerassi's retreat for artists in California. While there Hoffmann met an artist, Vivian Torrence, who also had a fellowship to work at the retreat. They decided to collaborate on a joint work dealing with the beauty and mystery of chemistry, combining Vivian's images with Roald's writing. The result was a recent book, *Chemistry Imagined*.[388] When Vivian planned a trip to Europe with a friend to gather material for the book, Roald contacted Notker as a former postdoctoral co-worker to ask him if he would escort the women around Munich. Notker did so and actually spent more time with them than he had actually planned. Vivian and Notker married in 1994.

The cerocene story is not yet completed. Other derivatives have been prepared by us[386] and others[389], and the subject has been reviewed recently by Fischer.[390] Recent calculations[391,392] have been interpreted by a model involving Ce(III) associated with the rings as a $(C_8H_8)_2^{3-}$ complex rather than as a Ce(IV) compound.[393] I find it difficult to see why such a model should be unique to cerium. For example, the next complex along the lanthanide series, $Pr(C_8H_8)_2^-$, cannot be oxidized to a cerocene analog. The essentially normal NMR spectrum of cerocene is also difficult to understand by such a model. Moreover, the photoelectron spectrum of cerocene is similar to that of thorocene and shows no f-electron band such as that in uranocene. If the interpretation of cerocene as a Ce(III) compound is correct, these aspects of experimental chemistry still need to be rationalized by theoretical studies. The differences may actually be a matter of semantics. The simple model in Figure 31B implies a shift of electron density from the cyclooctatetraene dianion rings to the central metal; thus, the metal is less like $Ce^{4+}$, and the rings have less charge than two dianions.

## Structures and Hydrolysis

Uranocene hydrolyzes slowly when treated with water; a THF solution 1 M in water has a half-life of about 1 day. Thorocene hydrolyzes much more rapidly. Kinetic studies of the hydrolysis of uranocene were started by an undergraduate, Roger Walker, and continued by several graduate students. Their work showed that electron-attracting ring substituents greatly accelerate the hydrolysis rate. For example, in a study of substituted aryluranocenes, Bob Moore found a Hammett correlation with $\rho = +2.12$, indicating a substantial rate-acceleration effect of electron-attracting groups.[394] If hydrolysis involves a rate-determining protonation of the ring, one would expect electron-attracting groups, which stabilize carbanions, to hinder reaction. This result suggests prior coordination of water with the metal; U(IV) compounds are generally strong Lewis acids, and such coordination would be enhanced by electron-attracting substituents. Deuterium

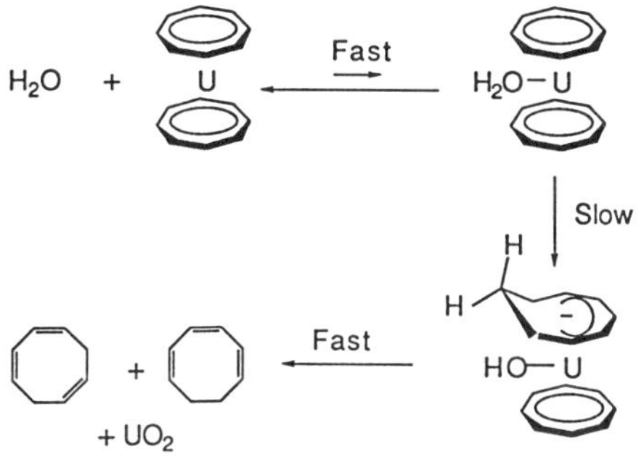

Scheme XI.

oxide reacts much slower than water; $k_H/k_D = 12$ for uranocene at 25 °C. Thus, ring protonation rather than coordination with the metal is the rate-determining step. These results were used to formulate the mechanism in Scheme XI. The essential feature is that a small amount of prior coordination occurs rapidly, followed by the rate-determining endo-protonation of the ring. The kinetics then follows the expression: $k_{expl} = K_{equilibrium}k_{protonation}$. Electron-attracting substituents would then have a larger effect in increasing $K_{equilibrium}$ than on diminishing $k_{protonation}$.

Having a variety of structures derived from cyclooctatetraene on hand, we could compare their relative hydrolytic stabilities. Treatment of cyclooctatetraene dianion with water results in instant hydrolysis to give a mixture of 1,3,5- (**36**) and 1,3,6-cyclooctatrienes (**37**) (eq 28).

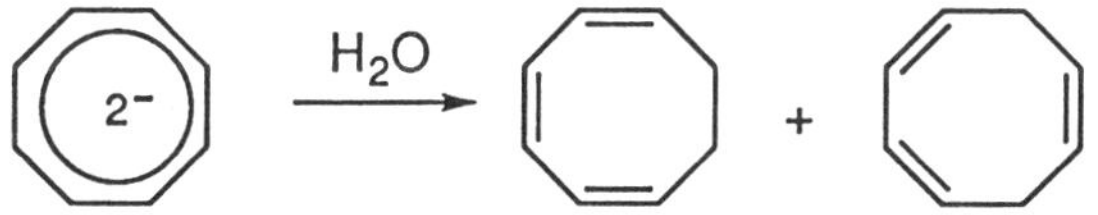

(28)

The ratio of these products depends on the gegenion, as shown by some examples in Table VII. The interpretation of these results is complicated by the fact that two rings are involved in different stages, but the product ratio is determined in each case by the second protonation. The variation found suggests strongly that the mechanism in Scheme XI applies generally; that is, that water coordinates first with the metal cation and then protonates the ring. This mechanism makes sense in several ways. Protonation of

Table VII. Hydrolysis Products of Different [8]Annulene Compounds in 1 M $H_2O$–Tetrahydrofuran

| Compound | % 1,3,5 (36) | % 1,3,6 (37) |
|---|---|---|
| $K_2COT$ | 76 | 24 |
| $Li_2COT$ | 57 | 43 |
| $Ce(COT)_2$ | 62 | 38 |
| $KSm(COT)_2$ | 56 | 44 |
| $KYb(COT)_2$ | 60 | 40 |
| $K_2Yb(COT)_2$ | 55 | 45 |
| $Th(COT)_2$ | 49 | 51 |
| $ThCl_2COT$ | 48 | 52 |
| $U(COT)_2$ | 36 | 64 |
| $UCl_2COT$ | 48 | 52 |

NOTE: COT is cyclooctatetraene.
SOURCE: Data are from reference 394.

the ring exo by uncoordinated water would produce hydroxide ion separated from a metal cation by a hydrocarbon fragment and would clearly be disfavored in a nonpolar solvent. Coordination with the metal cation would enhance the acidity of the water and facilitate the protonation step. Thus, this type of mechanism must have wide generality.

The large number of X-ray structures of organometallic cyclooctatetraene compounds now available, many from our laboratory, has also led to some important generalizations. Ionic radii provide a useful concept in inorganic chemistry. The effective ionic radius of an ion depends on its coordination number; in general, the more anions surrounding a cation in a crystal, the greater the effective ionic radius of the cation.[395] Many of the f-element organometallic compounds of cyclooctatetraene show this generalization, but with some notable exceptions. Table VIII gives some structural data for several potassium salts of complex ions containing cyclooctatetraene dianion.

All of the complex salts are of the type 38 with a trivalent actinide or lanthanide sandwich of two cyclooctatetraene dianion rings. The lower Yb–ring distance compared to the Ce–ring distance reflects the lanthanide contraction: as one goes along the lanthanide series from cerium to lutetium the effective ionic radius of the lanthanide becomes smaller. The potassium–ring distances show some variation but are all much larger than that in

# A Lifetime of Synergy with Theory and Experiment

Table VIII. Structures of Several Potassium Salts

| Compound | M | M–C (Å) |
|---|---|---|
| [K(diglyme)][U(MeC$_8$H$_7$)$_2$] | U$^{3+}$ | 2.732, 2.707 |
|  | K$^+$ | 3.263 |
| [K(diglyme)][Ce(C$_8$H$_8$)$_2$] | Ce$^{3+}$ | 2.746, 2.733 |
|  | K$^+$ | 3.166 |
| [K(diglyme)][Yb(C$_8$H$_8$)$_2$] | Yb$^{3+}$ | 2.610, 2.598 |
|  | K$^+$ | 3.191 |
| [K(diglyme)]$_2$(Me$_4$C$_8$H$_4$) | K$^+$ | 3.003 |

SOURCE: Data are from reference 375.

the dipotassium salt of a cyclooctatetraene dianion symbolized by **39**. In all of these cases the coordination around the potassium is identical: the three oxygens of a diglyme and the eight carbons of a planar [8]annulene ring.

The differences can be rationalized on simple Coulombic grounds. In **38** the potassium is attracted to the dianion but repelled by the trivalent central metal. In **39** the repulsion is with a monocation; thus, in **39** both potassiums can get closer to the ring. A corollary of this simple electrostatic argument also accounts for the two ring–metal distances in the sandwiches. In all three cases, the central metal is closer to the ring not complexed to potassium. One ring is attracted to K$^+$ as well as to a M$^{3+}$ and therefore is not as close to the M$^{3+}$ as is the other ring. This simple example shows that in many cases, the increase in effective ionic radius of a cation with increasing number of ligand lone pairs probably results from the mutual repulsion among the lone pairs and that in some cases it is necessary to consider next-nearest neighbors and not just nearest neighbors in interpreting effective ionic radii. That is, the "ionic radius" is not an intrinsic property

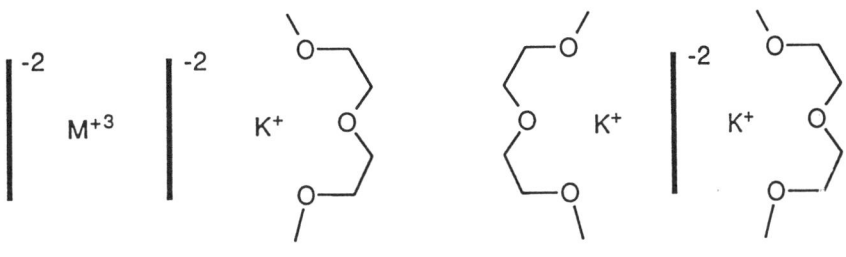

38                                              39

*A portrait of me in my office in 1984. Reproduced with permission from Lawrence Berkeley Laboratory.*

of the central cation but depends on electrostatic interactions among the ligands.

My experimental work with organoactinide chemistry occupied about 25 years but is now at an end. It was fun while it lasted, and it got me involved with a number of new techniques and ideas. I would sum up this work as having added to our knowledge of chemistry, especially in terms of bonding, structure, and reactions, but it has not led to important new materials or reagents. Useful materials were unlikely in any event because of the limitations imparted by radioactivity. In this regard, the lanthanides are undoubtedly far more important to organic chemistry, and I hope to continue some organolanthanide research. But understanding of organolanthanide chemistry is helped greatly by comparisons with organoactinide structures and reactions. This chemistry is not as rich as that of the d-transition series, but some of it has wide generality. Moreover, the story of my involvement in this chemistry as related in this chapter shows that organic chemists are not confined to the first few rows of the periodic table.

# Heterocycle Polycations

## "Densely Charged" Compounds

I recently found myself serendipitously in a new area of chemistry, an area of heterocyclic chemistry with compounds highly substituted with pyridinium and related cation groups. This story starts in 1980 when one of my graduate students, Kenneth Smith, was exploring "inverse sandwich" compounds, sandwich structures with two π-cation rings flanking a central dianion, symbolized as **40**.[396]

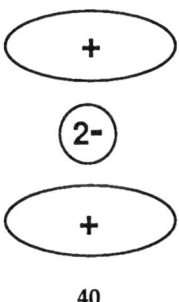

**40**

This type of compound was first alluded to years ago in Doering's seminal paper on the tropylium cation when he mentioned that tropylium sulfide does not have the possible sandwich structure (eq 29).[320]

$$\text{[bis(cycloheptatrienyl)sulfide]} \longleftarrow \text{[tropylium cation]} + S^{2-} \not\longrightarrow \text{[tropylium}^+ \cdots S^{2-} \cdots \text{tropylium}^+\text{]} \quad (29)$$

Ken Smith did not find any such structures either, but in a computational search for such compounds the cyclopropenyl cation ring system was an especially significant one. Triaminocyclopropenium cations are particularly stable cyclopropenyl cations and are readily available from the reaction of tetrachlorocyclopropene, **41**, with secondary amines (eq 30).[397]

$$\underset{\mathbf{41}}{\text{Cl}_2\text{C}=\text{C(Cl)}-\text{C(Cl)}_2} + R_2NH \longrightarrow (R_2N)_3C_3^+ \quad (30)$$

Ken had the idea of running a similar reaction with pyridine in hopes of getting the novel tetracation, **42**.

**42**

I thought that this was just too many cations close together, and I didn't give the reaction any chance of success. Nevertheless, like my esteemed colleague Ernest Eliel,[398] I will let good students try their own thing whenever feasible. Little did I realize that Ken would thrust me into novel and unanticipated chemistry.

Ken ran the reaction of pyridine with tetrachlorocyclopropene and obtained two unexpected indolizines, **43** and **44** (eq 31), reaction products I did not expect either.

# A Lifetime of Synergy with Theory and Experiment

(31)

Separating and purifying these saltlike compounds was not a trivial task, but Ken accomplished it and deduced the structures through analytical results and spectroscopy.[399] The reaction was further studied by Ken Waterman, and the structure of **43** was established by X-ray crystallography.[400] The structure of **44** was later confirmed by Drew Speer and Steve DiMagno[401] using NMR and X-ray crystallography. Some substituted pyridines were studied to show that the method has some generality. Preliminary kinetic studies suggested a reasonable reaction mechanism. This mechanism involves successive additions of pyridine to the strained cyclopropene double bond and elimination of chloride ion until a cyclopropyl anion is formed that is sufficiently long-lived to ring-open to the corresponding allyl anion. An electrocyclic ring closure followed by an elimination then gives the final product (Scheme XII). Formation of **43** or **44** depends on how many pyridines add before ring opening occurs. A transient color observed during the reaction could be the intermediate allyl anion.

In the mechanism in Scheme XII the pyridine has been added to the double bond in the correct order to end up with the observed products. That order is such that negative charge, where possible, is stabilized by an adjacent pyridinium cation and not in such a way as to put two pyridiniums on the same carbon if possible. Later work by Steve DiMagno found small amounts of the alternative isomers. The final step, the electrocyclic ring closure to the indolizine itself, has prior analogy.

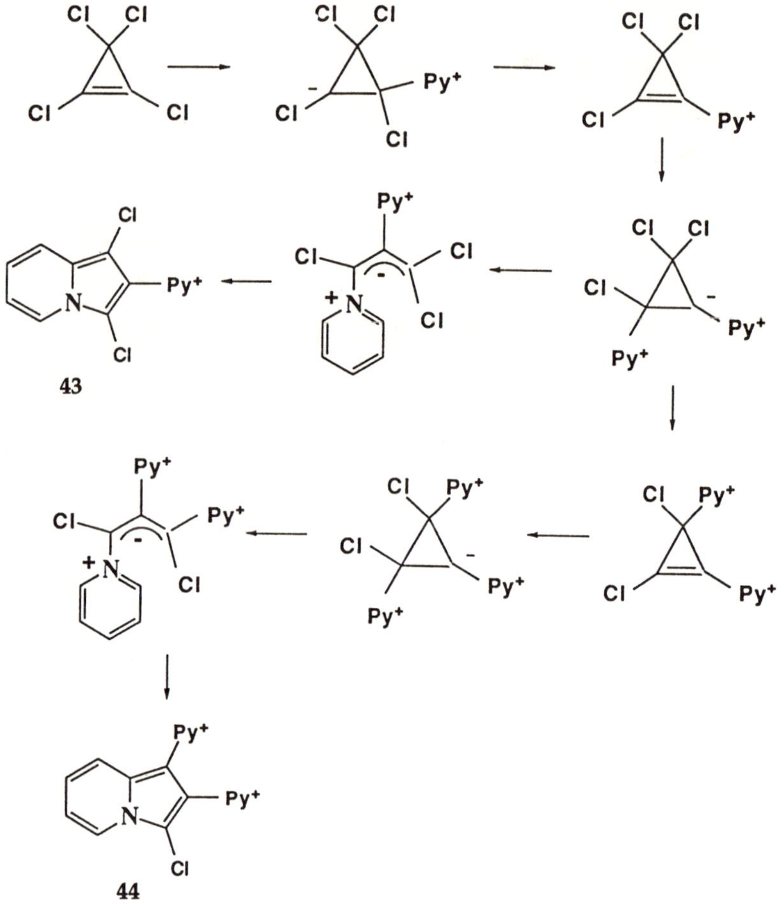

*Scheme XII.*

At this point the reaction has only limited interest. It makes an interesting mechanism problem for use on examinations. The important next step came when Ken Waterman studied the behavior of a highly nucleophilic pyridine, 4-(dimethylamino)-pyridine (DMAP).[402] The reaction of dimethylaminopyridine with tetrachlorocyclopropene gave a white crystalline material, a tetra-pyridinium substituted cyclopropene, 45, the first of a class of compounds that could be called "pyridiniumcarbons", in analogy to fluorocarbons and cyanocarbons. This is a rather stable material, much more stable than I would have expected for such a compound with its dense array of four positive charges. It is

## A Lifetime of Synergy with Theory and Experiment

amusing to note that this compound can be considered as the product of reaction with pyridine of **42**, the compound Ken Smith was trying for in his original reaction! Either by reaction of **45** with additional DMAP or by treating tetrachlorocyclopropene with excess DMAP, the ring-opened compound **46** was obtained as stable dark-red crystals (Scheme XIII), the second of our pyridiniumcarbons, an allyl anion completely substituted by pyridinium cation groups. This product, a positively charged allyl anion, will be referred to as an allylide.

These compounds are salts. Although the gegenions are not shown in the schemes, they are of course present. In many preparations the chlorides are produced but are frequently not easy to work with. These compounds cannot be handled as normal organic materials and have to be treated in many ways as inorganic salts. Purifications involve multiple precipitations or crystallizations, and extensive use of NMR was made for characterization as well as for establishing purity. Other techniques used were size-exclusion chromatography and electrophoresis. The

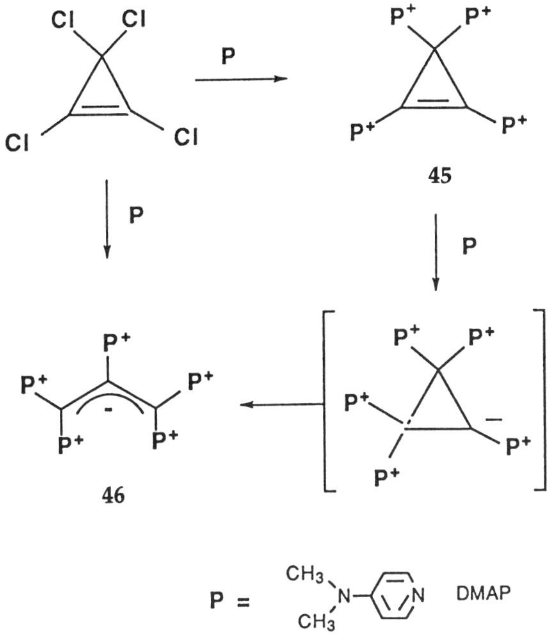

*Scheme XIII.*

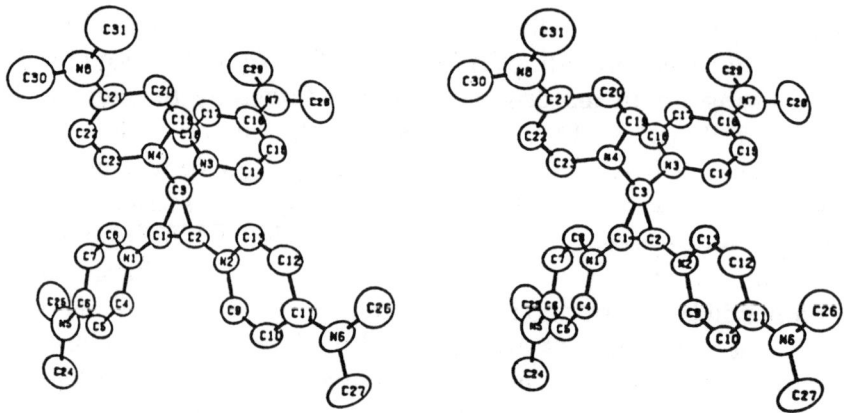

*Figure 33. X-ray structure of 1,2,3,3-tetrakis(4-(dimethylamino)pyridinium-1-yl)-cyclopropene, 45. This stereoscopic view shows the twisting of the pyridinium groups.*

gegenion anions could be readily replaced by using ion-exchange resins. The tetrafluoroborate and the hexafluorophosphate were especially useful because they tend to be less hydroscopic than the chlorides. The elimination of all traces of water from these compounds was always exceptionally difficult, and they often crystallized with varying amounts of water.[403]

The structure of **45** was established by X-ray crystallography (Figure 33).[404] In this structure it is noteworthy that the two pyridiniums conjugated to the double bond of the cyclopropene are almost coplanar with the double bond. The compound was found to react with various nucleophilic reagents. The reaction with dimethylaminopyridine is general for γ-aminopyridines. For example, the pyridine derivatives **47** and **48** gave mixtures of two allylides (Scheme XIV).

Less nucleophilic pyridines do not react normally. 4-*tert*-Butylpyridine, for example, gave only the allylide, **46**. The reaction is still not completely understood although it must involve reaction with the pyridine with liberation of DMAP, which then reacts with **45** to give **46**. Cyanide ion, however, does behave as a normal nucleophile, adding and ring-opening to give the corresponding cyanoallylide system, **49** (Scheme XV). Similarly, triphenylphosphine reacts with **45**; it adds, ring-opens, and gives

# A Lifetime of Synergy with Theory and Experiment

*Part of the research group in 1985. It was rare to get the whole group dressed up. The occasion was Ken Waterman's wedding celebration on the Berkeley campus. From left: Leyi Gong, Scott Gronert, John Rigsbee, Dave Eisenberg, Mike Kaufman, Drew Speer, Phil Sasse, Steve Bachrach, Bob Moore, Andy Koch, and Jin-Xiang Ni.*

*Scheme XIV.*

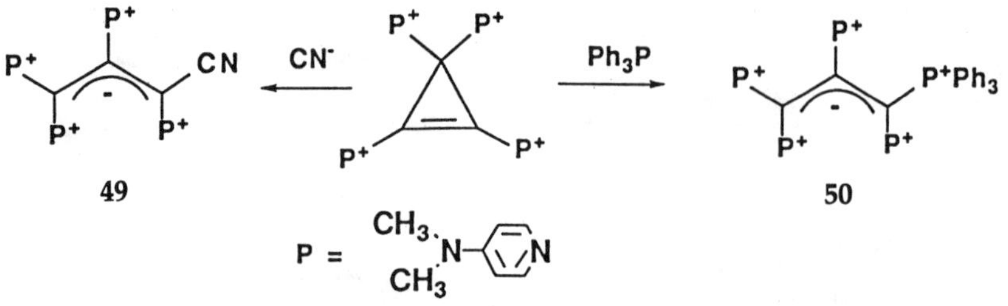

**Scheme XV.**

an allylide system that is a highly stabilized Wittig reagent, 50 (Scheme XV).

One of my last graduate students, Amy Feng, realized that the heterocyclic nitrogen in imidazole is further activated by the other nitrogen in the imidazole ring in the same manner that the γ-amino group in DMAP enhances the reactivity of the pyridine nitrogen. She found that indeed N-methylimidazole reacts as a normal nucleophile toward **45**; one mole adds and ring-opens to give the two cis–trans isomers of the imidazole-substituted allylide, **51**. Reaction of methylimidazole with tetrachlorocyclopropene gives the corresponding allylide penta-substituted with N-methylimidazole, **52** (Scheme XVI).[404]

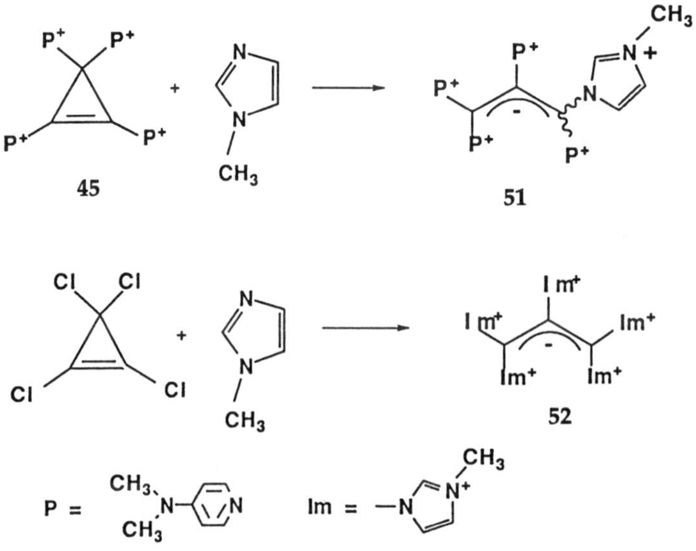

*Scheme XVI.*

Because these anions are all dark red and the conjugate acids that are formed on treatment with strong acid are colorless, they serve as their own indicators, and we could measure the p$K$ values from the spectra in buffered solution. In this way we have measured the p$K$ values of several of these compounds. The cyano compound **49**, for example, is an exceptionally stable anion whose conjugate acid has a p$K$ of –1. The other perpyridinium allylides have p$K$ values of about 3. That is, these anions are less basic than acetate ion. For comparison, propene itself has a p$K$ in the middle 40s, and thus substitution with five dimethylaminopyridinium cations decreases the p$K$ of propene by about 40 units or about 8 units per pyridinium group.

The tetrapyridiniumcyclopropene **45** reacts slowly in aqueous solution to hydrolyze with ring opening to give the bis-pyridinium acrylic acid **53**. The reaction is too slow to be preparatively useful, but Andy Koch discovered that a much better route is the reaction with methanol containing some formic acid (Scheme XVII).[405] The methyl ester **54** is accompanied by varying amounts of the free acid because it is difficult to keep all traces of water out of the mixture. The structure of this ester was also established by X-ray crystallography, and it has the typical kind of structure one would expect. The external carbon–nitrogen bond distance to the dimethylamino group of about 1.32 Å is close to that found in the tetrapyridiniumcyclopropene structure, **45**. In this case, however, the two rings are twisted with respect to each other, reducing conjugation to the central double bond.

We tried various reactions of this ester. Jung and Buszek have shown that a monopyridinium acrylic ester is a good Diels–Alder dienophile that reacts readily with dienes.[406] We have not been able to get the bis-pyridinium compound to undergo any

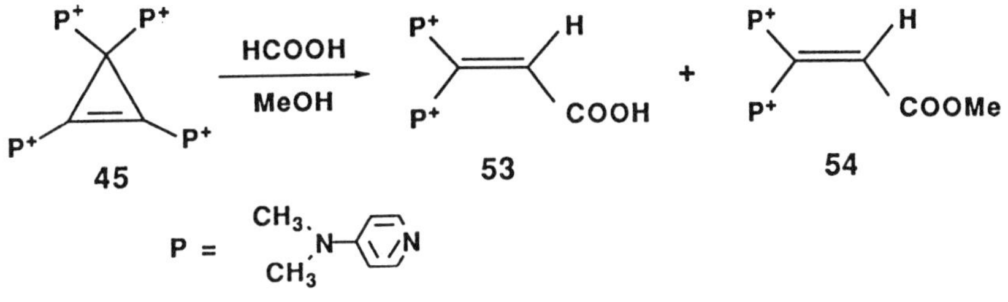

Scheme XVII.

similar reaction. Apparently the steric hindrance involved with the two twisted pyridinium rings is just too great. We tried reactions under high pressure making use of Bill Dauben's high-pressure apparatus but without success. The ester can be hydrolyzed to the acid, and the ionization of the acid to the anion can all be shown by small changes in their UV spectra. From these differences it was possible to derive a pK value for the acid of about 1.3; thus, 53 is a rather strong acid undoubtedly because of the inductive effect of the two pyridinium cation groups.

The reaction of dimethylaminopyridine with tetrachlorocyclopropene is so facile that we explored comparable reactions with other highly halogenated compounds. Hexachlorocyclopentadiene reacts to give stable cyclopentadienyl allylides (Scheme XVIII). The completely substituted pentapyridinium cyclopentadienyl allylide, 55, is another pyridiniumcarbon.[407]

The reaction itself is not straightforward because an overall reduction must occur to generate these products; the overall stoichiometry is unknown. The pentapyridinium cyclopentadienyl species 55 is a stable molecule and a rather weak base. It does not react with ferrous ion, and in fact we have not been able to make any metallocene derivatives from it. It does become protonated in strong sulfuric acid, and from the NMR spectra of sulfuric acid solutions we could deduce that the pK is about –6. That pK can be compared with the pK of 10 for monopyridinium cyclopentadienyl anion shown a number of years ago by Kosower and Ramsey.[408] Cyclopentadiene has a pK in water of 16,[329] which thus indicates that a single pyridinium ring increases the acidity of cyclopentadiene by about 6 units. This number is somewhat less than the average value of about 8 units that we found previously for the effect of five pyridinium rings on the acidity of prop-

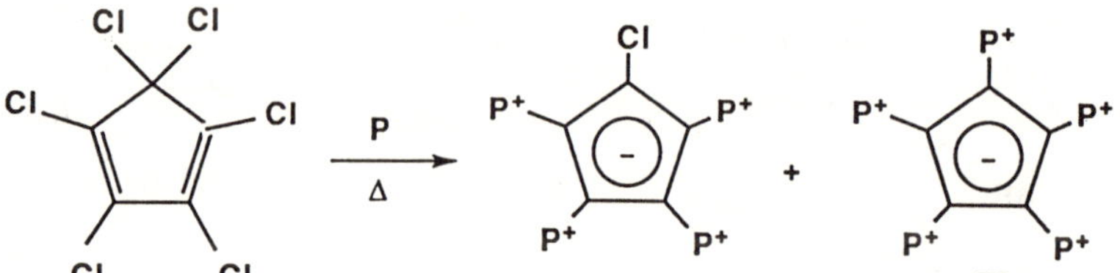

*Scheme XVIII.*

ylene. Moreover, if we compare the p$K$ of 16 with our p$K$ of –6 for the pentapyridinium system, **55**, that means that five pyridinium cations reduce the p$K$ of cyclopentadiene by about 22 p$K$ units or about 4½ units per pyridinium ring. This number is less than that for a single pyridinium group probably because of a saturation effect and because the conjugation of the dimethylamino groups in **55** puts the positive charge effectively somewhat farther away from the anionic charge and reduces the stabilization accordingly.

The reaction of nucleophilic heterocycles with highly halogenated alkenes and aromatics appears to be rather general. Several examples were demonstrated such as the reaction of DMAP with hexafluorobenzene to yield two new compounds, **56** and **57** (eq 32).[407]

(32)

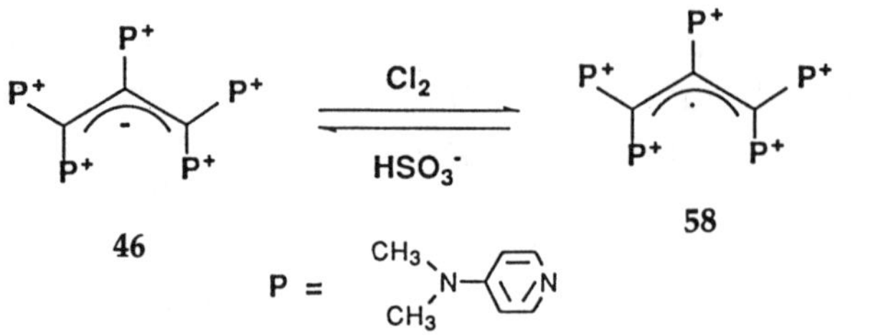

Scheme XIX.

The allylides undergo reversible one-electron oxidations on cyclic voltammetry at relatively low potentials; **46**, for example has $E° = 0.44$ V vs. SCE.[404] Chemical oxidation produces radicals that are remarkably stable. They can be reduced back to the corresponding allylide. The radical **58** produced by careful oxidation of **46** with chlorine (Scheme XIX) was ion-exchanged to the hexafluorophosphate salt, which gave crystals suitable for X-ray crystallography.[409] The structure (Figure 34) shows highly twisted pyridinium rings but an allylic bond angle of 126°, similar to the value of 124° found by electron diffraction of allyl radical itself.[410]

By using a well-known thermodynamic cycle, the acidity of the conjugate acid **59** of the allylide **46** and the reversible oxida-

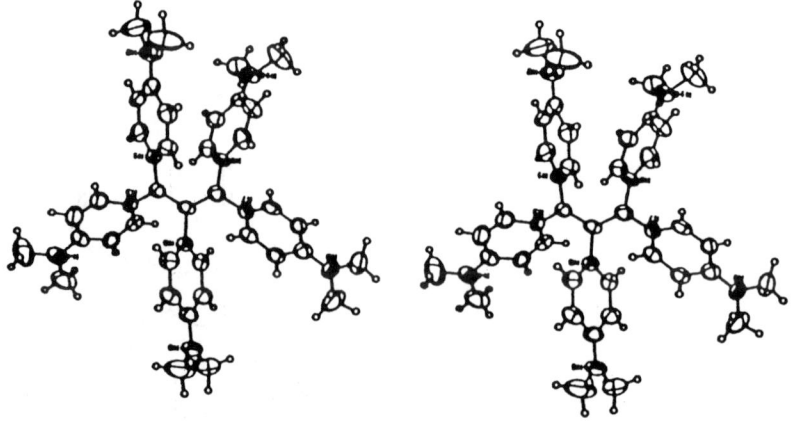

Figure 34. X-ray crystal structure of 1,1,2,3,3-pentakis-(4-(dimethylamino)-pyridinium-1-yl)allyl radical hexafluorophosphate, **58**. This stereoscopic view also shows the severe twisting of the pyridinium groups. (Reproduced from reference 409. Copyright 1991 American Chemical Society.)

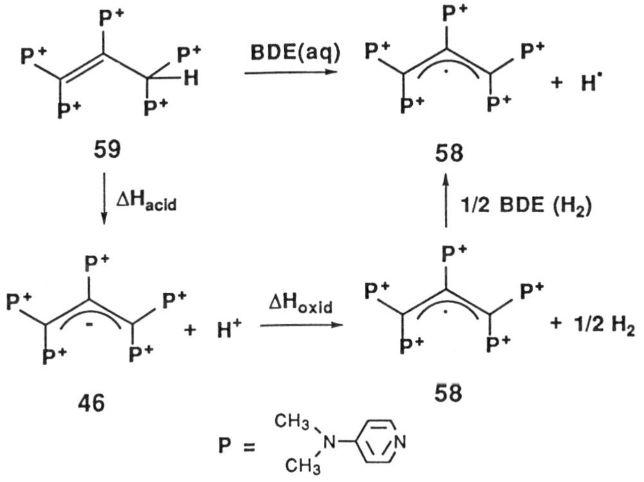

Scheme XX.

tion potential of **46** gives the bond dissociation energy (BDE). This cycle is shown in Scheme XX. Because ½BDE of $H_2$ is 52.1 kcal/mol, the BDE of **59** is given as BDE = $\Delta H_{acid} + \Delta H_{oxid} + 52.1$. We equate $\Delta H_{acid}$ with $\Delta G°_{acid}$ from the pK value and derive the corresponding BDE as 72.0 kcal/mol.[411] This value is lower than propylene itself (86 kcal/mol) undoubtedly because of the extended conjugation in the polypyridinium system.

Steve DiMagno thought of attaching long alkyl chains to the allylide to decrease the water solubility. He did this by replacing one methyl group in DMAP by a decyl, hexadecyl, or docosyl group. The resulting allylides are soluble in chloroform, and these solutions were used to coat electrodes. The treated electrodes were found to have rapid electrode kinetics and specificity for electroactive anions.[412]

Thus, the simple experiment that Ken Smith ran and that did not work as he had hoped instead produced a rich harvest of unexpected chemistry. Moreover, all of the succeeding work followed because he did not simply throw away a failed reaction mixture but instead worked up and characterized the unanticipated products. All of this chemistry was the outgrowth of a completely different project, the attempted preparation of inverse sandwiches. I would not have thought that compounds so highly substituted with cationic groups would be so stable and so easy to produce. These novel compounds should have potential as interesting and useful materials. The indolizines, for example, show a dependence of fluorescence maxima on concentration, a phenomenon indicative of a stacking aggregation.[413] We had hoped to produce polycationic polymers that might have properties of interest, but we could not get high-molecular-weight materials. In the polymer and electrochemical studies we benefited greatly by collaboration with my Chemistry Department colleague, Professor Jack Porter, an expert in these areas. We had obtained a NSF grant jointly to support the initial research, as well as some of his work with polyacetylene. Unfortunately, all attempts to renew this grant or find other funding failed. Although these heterocyclic cations are unusual compounds, I was not able to establish specific uses as materials at this stage. Accordingly, this project has now been phased out, but we did have a decade of fun chemistry. Fortunately, however, this chemistry is not at an end. It is being continued by Professor Robert Weiss, Institute of Organic Chemistry, University of Erlangen, Nürnberg, Germany, who independently discovered some of this same chamistry. His group has, for example, prepared the fully pyridinium-substituted benzene analogous to 57 by a clever reaction strategy and has found it to have unusual ion pairing properties that should indeed lead to novel materials.[414]

# Odds and Ends

## Organic Plasma Chemistry

In 1961, I attended a physical chemistry seminar on plasmas generated by microwave discharge and wondered what such conditions would do to organic compounds. Hal Ward, an NSF Postdoctoral Fellow at that time, took up the challenge. He set up a magnetometer microwave discharge apparatus and studied the behavior of a number of compounds passed through in a stream of helium. In general, aliphatic compounds gave low conversions to mixtures of many compounds, whereas aromatic compounds tended to give fewer products with higher conversions. Toluene, for example, produced varying amounts of tar and gave a substantial amount of benzene, but we were surprised by the formation of phenylacetylene, styrene, and ethylbenzene.[415] The study was continued by another postdoctoral co-worker, Adam Heller, and two graduate students, Steve Rodemeyer and Bob Bittman. A number of various substrates were studied, including benzaldehyde, aromatic ketones, and several heterocycles. Different carrier gases were used, and reaction mechanism studies were carried out with isotopically labeled compounds and carrier gases. Deep-seated rearrangements were found that were indicative of complex reactions and mechanisms. For example, the use of toluene-7-$^{14}C$ gave benzene that contained the label.[416] Pyridine gave as principal products benzene, toluene, acetonitrile, benzonitrile, and HCN.[417] No useful chemistry emerged from these studies, and most of this work was never published.

This work shows that not all interesting or intriguing questions will lead to significant or useful new chemistry. Timing is also important. It is possible that further study with plasmas might still lead to something useful, but that was not evident from our results and along the way one must always weigh the potential value of further investment. This balancing of effort invested versus potential yield is just as true in academe as in industry. The only difference is that industry needs ultimately to get a saleable product; the academic scientist searches instead for fundamental principles. In the present case, for example, the behavior of organic compounds in a plasma might have important relevance to organic chemistry in outer space, but that question is a different one that would require different techniques and experiments.

## Textbook: *Introduction to Organic Chemistry*

In addition to the course in physical organic chemistry, my other principal teaching assignment over my four-decade career were the lectures in the "major's" organic chemistry, the year course in introductory organic chemistry taken generally in the sophomore year. I tried many textbooks during my first two decades in Berkeley trying to find the ideal one to teach from. I failed. Each text I used had major problems, and I was finally persuaded to write one myself. But not alone. I wanted a coauthor with whom I could test ideas and writing and to share the work involved even though I realized that we would each be doing three-quarters of the work. Fortunately, Clayton Heathcock agreed to be a coauthor.

Our original organic group of Calvin, Cason, Dauben, Jensen, Noyce, Rapoport, and me are now all emeritus or deceased (Jensen), although Dauben, Rapoport, and I are still active professionally and in research, and Calvin comes regularly to seminars and meetings. Of our present full Professors in organic chemistry, Heathcock came in 1964, followed by Paul Bartlett in 1973, Peter Vollhardt in 1974, Bob Bergman in 1977, and Peter Schultz in 1985. This strong group will certainly maintain the traditions of Berkeley organic chemistry into the new century.

# A Lifetime of Synergy with Theory and Experiment

*Sue's parents owned a cabin in Island Park, ID. Clayton Heathcock (left) and his family visited us there in the summer of 1970. Fly-fishing is one of the special treats of that area, as this picture of us indicates. We are celebrating our fishing success with ice cream cones.*

Clayton Heathcock and I started writing our textbook in the early 1970s. We made a good combination with my abilities in theory and physical organic chemistry and Clayton's strengths in synthetic chemistry. We share many teaching philosophies and had no major arguments or disagreements. For example, we both agree in an emphasis on the experiment, and thus many of the examples of organic reactions used in the text came from the well worked out procedures in *Organic Syntheses* and were given with actual yields and frequently with a summary of experimental conditions. We also agree in trying to explain concepts and chemistry without lying to students or talking down to them. I believe that these explanations in terms of fundamentals are a unique strength of the textbook. We are amused to find some of these explanations, even verbatim, used in subsequent textbooks by others.

We each wrote the first drafts of individual chapters, which the other then edited and rewrote. Clayton was especially good in getting my writing down to a sophomore level. The result was also to even out our individual idiosyncrasies so that many people have told us they could not tell who wrote which chapter. During part of this time, Clayton lived across the bay in Marin County. I live much closer to campus. Much of our discussion and writing together was done during evenings at my house, and many's the night that Clayton slept on our couch rather than face the long drive home and back the next morning.

The book was typed by one of Clayton's daughters and by my secretary, Lynne Gloria. Lynne's time was shared with Henry Rapoport; her office adjoined both of ours. She came to us as a young woman and stayed with us for more than 25 years. Lynne learned rapidly and was a very capable secretary. She left only because she had reached the maximum payroll classification and salary as a professorial secretary. To increase her salary she had to move to an institute. Our university, like most organizations, pays personnel on the basis of the number of people they supervise rather than on the quality of their work. My wife Sue and son David prepared the index to help our students use and test the manuscript. Clayton and I arranged to provide copies of the manuscript for use in our courses so that we could test it in the classroom. David had recently taken the course and had not forgotten his organic chemistry. He marked index entries, and Sue keypunched these entries onto IBM cards. These were then converted into a printed index with a program written by a graduate student, Glen Toczko, for use on the chemistry department computer, an incredibly primitive computer by modern standards.

Our textbook, *Introduction to Organic Chemistry*, was published in 1976. It was successful and is now in its fourth edition. Earlier editions were translated into German, Japanese, French,

# A Lifetime of Synergy with Theory and Experiment

*Lynne Gloria in 1968. Lynne was a secretary for me and Professor Henry Rapoport for more than 25 years. She recently retired from the University.*

Portuguese, and Spanish. For the fourth edition, published in 1992, we added a third coauthor, my former high school colleague Ed Kosower, in order to add expertise in bioorganic and biophysical organic chemistry. These topics have become increasingly important in today's organic chemistry even at the undergraduate level. On the other hand, I believe that our textbook contains too much theory. We had to include many aspects of MO theory (e.g., the Woodward–Hoffmann rules) because the market requires these topics, but I do not think they belong in a sophomore textbook. Indeed, I think there is too much theory in freshman chemistry. Most students simply do not need MO theory at that level. The chemistry majors, who do need it, should get it later when they already know some chemistry.

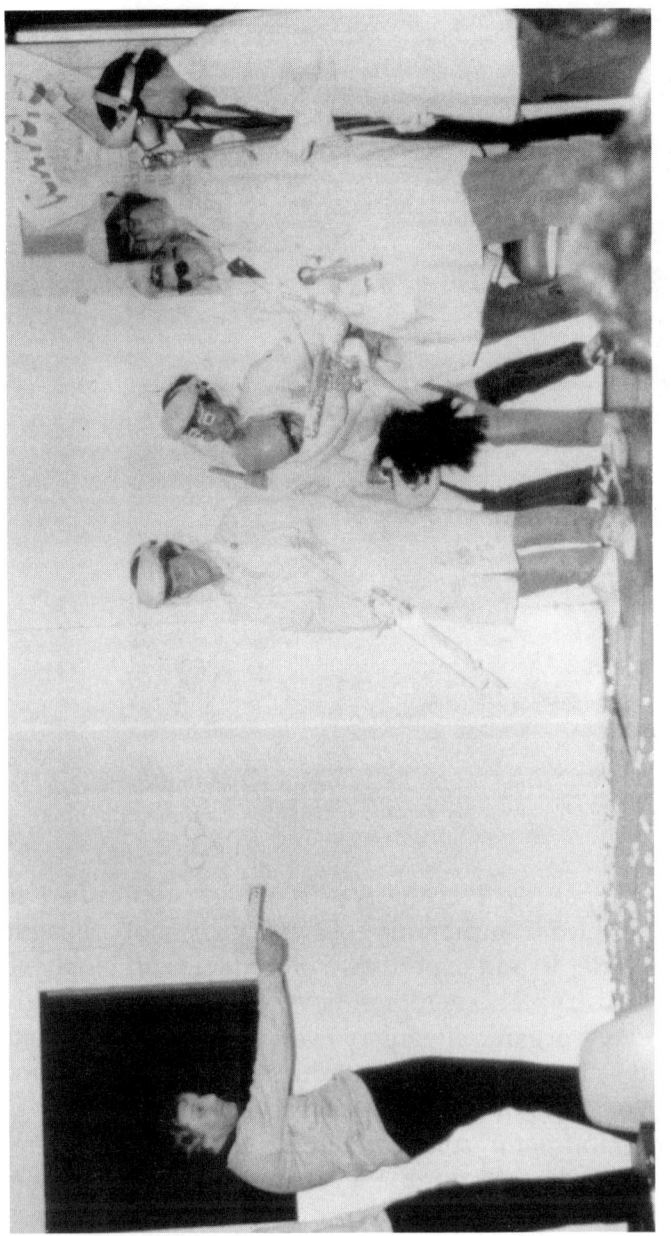

The research students and faculty of our chemistry department have a long tradition of celebrating the Christmas holidays with a party and skits. I have frequently been called upon to perform in these skits playing such roles as Siskel (to Heathcock's Ebert) and Columbo. In this 1985 skit, I played Captain Kirk repelling a group of alien invaders. My shot ricocheted off the bald head of Bob Bergman. The shot hit Clayton Heathcock sitting in the front row, who then slumped forward and lay prone on the floor. My next line was "My God! There goes the third edition!"

I am already thinking about the next edition. Some MO theory will probably have to be retained, but I hope that it will be possible to include it or not, depending on the individual teacher's approach. This type of choice may become especially feasible through textbooks written for computers. The new age of computers is bound to have a major effect on teaching and may increase the flexibility in using textbook materials. At such time that all students have personal computers, will conventional textbooks still have a role? In fact, will chemistry still be taught in large halls with a single teacher lecturing to hundreds of students? The coming decade should be an interesting one for the teaching profession. Even though I am now emeritus, I hope I will play some role in the changes to come.

## *Journal of Organic Chemistry*

In 1989, Clayton Heathcock became Editor-in-Chief of the *Journal of Organic Chemistry,* and I was appointed as one of his associate editors. It was expected that I would handle manuscripts in physical organic and theoretical chemistry. Almost immediately, the number of theory manuscripts submitted to the journal increased substantially. These papers were quite uneven in quality, and I found it necessary to publish an editorial[418] detailing what types of theory papers the journal would accept. The editorial had the desired effect, and most of the recent papers dealing with theory are good. One of the interesting changes in recent organic chemistry is the increasing role of theory in papers dealing primarily with experimental and synthetic organic chemistry. Molecular mechanics calculations are now frequently included to bolster stereochemical arguments based on conformations or to compare NMR coupling constants. Semiempirical calculations of the AM1 type are also common; I'm sure it is a source of great joy to Michael Dewar that his offspring has become such a useful tool to organic chemists. I have handled a large number of papers focussing on ab initio calculations, and a number of papers have such computations included among experimental work.

In this way I believe that my editorship has had an impact on the literature of organic chemistry. We all have a responsibility to our profession. Being an editor is time-consuming and is work, but it has its rewards, and I feel I am making a contribution. One thing that has struck me in this role is how many chemists feel the same way. Quality journals cannot exist without many responsible referees. Reviewing manuscripts is time-consuming and unsung, yet most chemists take this responsibility seriously and conscientiously. I appreciate their efforts, and I thank them. At the same time, there are those chemists, including some with famous names, who are irresponsible. Some reviewers routinely recommend "Publish without change" without further comments, when other reviewers of the same manuscripts find numerous errors and deficiencies. Other reviewers hold manuscripts for many weeks despite repeated attempts by the staff to elicit replies. Of course, we have come to know who the irresponsible referees are and have stopped using them. But I wish it were possible to report a reviewer's "grade" to promotion or award committees! I can only urge all chemists to take their reviewing responsibilities seriously and to provide helpful and timely reviews with suitable comments. Such critical reviews of the papers of others are as much our responsibility as writing our own papers; they help to keep the chemical literature of high quality and understandable and reproducible by others.

One other comment in this regard is that most ACS journals are international rather than national in scope. Many papers submitted to *JOC* come from foreign laboratories and are written by authors whose native language is not English. The language is frequently not good English. I recognize how difficult it must be for one whose English was learned later in life to write well in such a difficult language, but English is now the universal language of science, and it is important that the science be expressed clearly and accurately. Although some commercial editorial services are available for this purpose, it is clearly desirable for such an editor to know chemistry in order not to distort an author's meaning. One chemist now retired from industry has volunteered his services on occasion to *JOC* to help rewrite manuscripts in which the English is deficient. This is one way for a retiree to stay professionally active and thereby to provide a valuable service. It is a role that others may want to emulate.

A related point that has impressed me as an editor helping to produce the primary literature and as a writer of reviews who converts the primary literature to secondary literature is how the vastness of chemistry consists of so many individual contributions. Individual investigator chemistry is considered "small science" that has less impact on the public compared to the big machines of physics or the big telescopes of astronomy. But the whole of chemistry is big and, although there are unifying principles, these are based on a diverse and rich body to which a great many individuals have contributed excellent science. Each such contribution may be "small", but they mesh together to build and create a fabulous and incredibly rich whole.

An editor must also on occasion mediate controversies. I have mentioned several times in this story that chemistry is a complex subject that we make understandable with the help of simplifying concepts, but these concepts themselves can be difficult to define definitively. A physical observable, a scientific property that is the expectation value of an operator acting on a wave function, is not ambiguous and has reality; other properties might have rigorous definitions but are more or less ad hoc. Many of our chemical concepts, such as atomic charge, electronegativity, ionic character, steric hindrance, and bond order, are of the latter type. Discussions with these principles can involve controversies among individuals and give rise to polemical papers. Some controversies of this type might have little merit, but others can be useful and can help the science grow. The decision to publish such controversies must be made carefully and requires a degree of judgment; it is rarely a matter of being simply right or wrong.

An entirely different type of controversy arises when a paper is published that contains a significant error that has escaped the reviewers. Often, such cases are handled informally, and the author simply publishes a subsequent correction. In other cases, the error is too trivial to be of concern. But sometimes an author will make a serious error that he or she is unwilling to recognize and correct. This situation then stimulates others to submit manuscripts pointing out and correcting the error. Some journals refuse to publish papers in which one author reinterprets the data of another without adding new data. Yet, if the original paper is important and contains a significant error, some mechanism is required for a correction.

*JOC* has published such manuscripts, and I have been involved in some such controversies. This situation is the most

*Paul Schleyer (left) and Norman (Lou) Allinger during an editorial board meeting in 1994. Paul and Lou are the editors of the* Journal of Computational Chemistry. *I have been on the editorial board of this journal since its inception.*

time-consuming and difficult for an editor. One important question is immediately itself controversial: Should the original author be a referee for the "correction" paper? The original author might have cogent points to make, but such authors also tend to be quite defensive and no one likes to be corrected. In any event, reviewers are particularly important in such cases, and I have found it necessary to do extensive background reading myself to serve effectively in such adjudication.

## Ethics

The discussion of errors in the literature leads naturally to the matter of deliberate errors and thus of ethics. Scientific fraud is as old as science and has been much in the news recently, but fortunately it appears to be rare in the physical sciences. In large part, this relative rarity is simply because physical data are exact and can be more or less readily reproduced. In C. P. Snow's *The Affair*[419], the first indication of a scientific fraud comes from the attempt of another laboratory to repeat the experiment. Still, I have heard of occasional cases of students who have invented chemical data such as spectra. Research directors can often be subject to such fraud. I don't think that any of my students have defrauded me in this way, but it is certainly possible that there could have been such instances that I was not aware of. Eventually "the truth will out", especially for results that are timely and significant. The danger lies more in work of less significance that can still provide a dissertation or other immediate reward. Many results are not worth repeating and in fact are not repeated unless they are suspect or inconsistent with current thought.

The topic of ethics is not really discussed in my research group because it should not be necessary to remind anyone that one should not plagiarize or steal. The matter of intent, however, needs to be emphasized. I have known of several cases in which an idea was presented to another, the exchange was forgotten, and the idea surfaced later by the other person without attribution. I have done this once or twice myself, fortunately in only minor ways, because the original discussions simply vanished from my memory. Similarly, I have had discussions with others in which my ideas were then published later by them, and I am sure that the same thing happened with them. Humans are fallible, and memories are not always reliable. These incidents are not violations of ethics because the intent to defraud was absent. The only way to limit such incidents is to always take careful notes. Literate humans are able to supplement memory with a written record.

Another aspect of scientific ethics concerns the handling of experimental data. It goes without saying that one should not falsify data; this type of cheating is obvious. A grayer area is that of choosing only those data that fit one's theory. This gray area can be as varied as neglecting to cite those experiments that disagree with one's theory, or selecting those experimental points that make a good straight line. We recognize, of course, that one frequently has reasons for neglecting some data, but this decision clearly must be reached with a healthy dose of intellectual honesty. That is, one should cite those experiments that are inconsistent with the theory and present one's arguments for ignoring them; posterity will judge whether one was correct.

## Awards

The California Section Award in 1964 turned out to be "my first award and not my only award", to recall the earlier phrase. In 1967, I was awarded the Petroleum Award at the American Chemical Society meeting in Miami. Sue accompanied me on that trip, and we left the children with my parents in St. Petersberg. The occasion provided the opportunity for my parents to meet

Receiving the 1967 American Chemical Society Award in Petroleum Chemistry from Chet Warner (right) of the sponsoring company, Precision Scientific Corporation. This award was my first national ACS award.

Sue. My father had had a debilitating stroke in 1959, and they were not able to attend our wedding. A few months later, shortly before my father's 66th birthday, he suffered another stroke and was hospitalized. I flew east to visit him and the airport limousine was involved in an accident. I suffered only bruises but was taken to a hospital for observation—my father's hospital. By this circumstance I got to see him just before he died. He had been a heavy smoker for most of his life, even after being diagnosed with circulatory problems.

In 1969, I was elected to the National Academy of Sciences. This was a prestigious event, especially at the age of 41. The announcement came as a complete surprise and in humorous circumstances. I have told of the month I spent with Professor Charles Coulson in April, 1969, during my first "traveling Sabbatical". At the end of that time we returned home via a few days in Dublin. Sue, our daughter Susan, and I (David had stayed behind in Berkeley) had expected to fly home directly, and I stopped off at the TWA office across the street from our hotel to confirm our flight plans. I learned that the flights had changed, and we were scheduled to stay overnight at Shannon Airport. This change in plans angered me, and I returned to the hotel in a foul humor, stopping in the lobby to pick up a telegram before displaying my anger to my family. After expressing my wrath at this change in plans, I opened the telegram, which came from then Dean of the College of Chemistry, Hal Johnston, congratulating me on my election to the National Academy of Sciences. This notification of such a singular honor was so unexpected at that time, and so totally out of context with my mood, that we all had a good laugh. And, in the event, our overnight stay at Shannon turned out to be delightful.

The National Academy of Sciences is an honorific organization established by an act of the Congress in 1863.[420] At the time of my election there were fewer than 900 members representing all branches of science. Thus, election to the Academy is a distinctive honor. Even now, despite the expansion of numbers of scientists in the last few decades, the Academy numbers fewer than 1700 members.

I already mentioned the Humboldt Senior Scientist Award, which resulted in our spending the first half of 1976 in Germany. In 1978, we were invited to be guests of the Humboldt Foundation to help celebrate their 25th anniversary that December of the reestablishment of the Foundation. The occasion enabled us to spend Christmas in Munich to cap a most pleasant trip. I have served as a referee for the Foundation for many years.

In 1982, I was awarded the Norrish Award in Physical Organic Chemistry sponsored by the Northeast Section of the

*In 1982, I was the Raymond and Beverly Sackler Distinguished Lecturer in Chemistry. Sue and I spent a delightful week in Israel that October when I lectured at several universities.*

American Chemical Society, and in 1989 I received a Cope Scholar Award. In 1993, the Bavarian Academy of Sciences at their February meeting elected me as a corresponding member. This recognition makes me especially happy because my father as a native Bavarian would have greatly appreciated this honor to the Streitwieser name.

A local honor was accorded me quite by surprise during my first year as emeritus, 1993. I attended Commencement that summer in part because it would be my first in my new status and in part to present Amy Feng, one of my last graduate students, with her diploma. On the occasion I received the Berkeley Citation, a certificate for services to the Berkeley campus. I learned later that it was Amy's job to ensure that I would be present at that commencement!

## Hobbies

Many chemists I know have a single-minded devotion to chemistry. Although I love chemistry, it has not been a monogamous love. I also enjoy doing other things, such as flyfishing. Flyfishing requires one's complete attention. One cannot think about chemistry and expect to flyfish successfully. I fish primarily in the area where Idaho, Montana, and Wyoming meet. This area is beautiful country, and it is a joy to stand in waters, some calm and some swift, surrounded by trees and meadows and expansive skies, even if I cannot think about chemistry at the same time. My fishing areas are often shared with wildlife—some, like moose, not too closely! I will never be a really good fisherman because I do not spend enough time at it, but I am good enough to catch fish and have fun.

I have always enjoyed photography but did little of it in my early Berkeley years. In 1971, during my first trip to Japan, I bought my first Nikon camera with a series of lenses. This Nikon was my first good camera, and I have taken many pictures ever since. I joined the Berkeley Camera Club in 1973 and have fun competing in their slide divisions. I have learned a great deal about photography during this period and am capable of taking some good images. My greatest success so far is a slide of a raft in

Picture of me taken unawares on the South America tour I took in 1994 to see the solar eclipse. (Photograph taken by Diane Conner.)

the rapids of the Colorado River that was "Slide of the Year" in the Northern California Council of Camera Clubs in 1990. In recent years I have taken some trips that have had no chemistry connection at all. I saw my first total eclipse of the sun near Oaxaca, Mexico, in July, 1991. That was such a thrilling sight that I traveled to South America to see the solar eclipse of November 3, 1994, near Iguazu Falls. Our group was again fortunate to have good weather, and I repeated the thrilling experience of the total eclipse. I was also delighted to obtain some excellent photographic images of that eclipse.

The San Francisco Bay Area has many amenities, and I've greatly enjoyed the opportunity to live here all these years. California has lost some of its appeal through earthquakes, fire, and population growth, and the Bay Area has its share of these problems. Nevertheless, many students come here and do not want to leave, although many of the jobs in chemistry are in other parts of the country. I've always counseled them that no place is perfect and every place has something to recommend it; one should take advantage of what an area has to offer. One of the things the Bay area offers us is opera, but it took me a few years to find it. As I mentioned already, Sue and I have shared Opera early in our relationship and now go to performances of the San Fran-

# A Lifetime of Synergy with Theory and Experiment 253

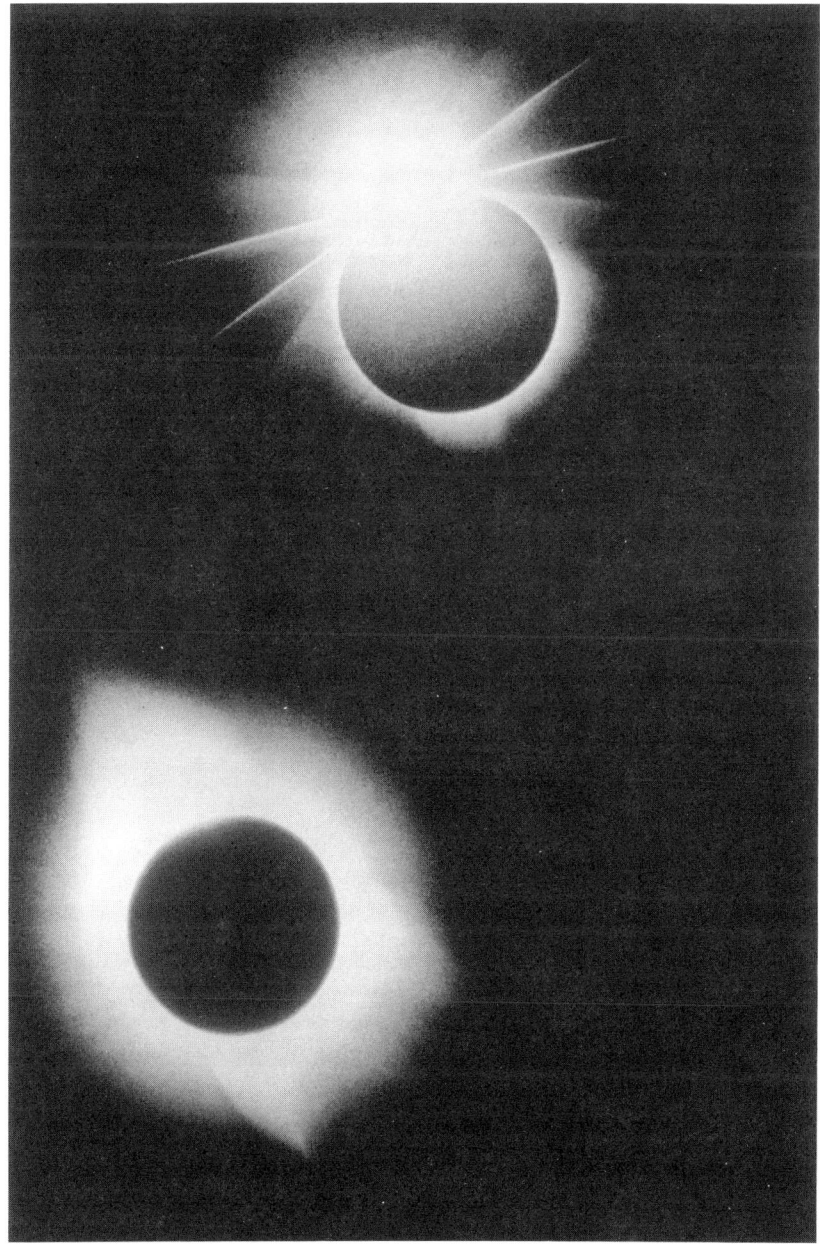

*Two views of the eclipse of November 3, 1994, showing the diamond ring effect and the corona. This slide was a "Nature" award winner at the Berkeley Camera Club. (Reproduced with permission from A. Streitwieser. © A. Streitwieser.)*

Gubbian Lock, a prize-winning slide in a competition at the Berkeley Camera Club. The picture was taken in Gubbio, Italy.

cisco Opera Company every season. We celebrated our 25th wedding anniversary in 1992 with a pilgrimage to Bayreuth, to see the Ring Cycle at that Mecca for Wagnerians. I've always enjoyed music and have regretted never having had musical training in my youth, a consequence of early poverty. I bought many recordings in high school only to have them superseded by LPs after the war. Now, of course, we have a large CD and video collection of opera and other musical performances.

Another amenity of the San Francisco area is good food, not only in restaurants but in markets. Sue always had a talent and desire for cooking, and during our marriage she has taken many classes in a variety of cuisines. We actually do not go out much for "fine dining" because we dine so well at home. We also complement her dinners with appropriate wines. I started a wine cellar in 1965 with the purchase of two cases of Bordeaux and have built up a "steady state" cellar of about $10^3$ bottles. In 1969, we held a wine tasting with a group of friends. We served several wines of the same type in bags for a blind tasting. Each of us ranked the wines by personal preference, and at the end we compared notes and then revealed the wines. This tasting was fun, and another member of the group hosted one a few months later. The host buys the baguettes and cheese, and we share the cost of the wines. We have held such wine tastings 6–8 times a year ever since. We celebrated significant anniversaries with special dinners or outings. The year 1994 was the 25th anniversary of this wine-tasting group, and 12 of us met for a week together in Bordeaux to honor the occasion. The tastings at famous chateaux and our fabulous dining experiences were arranged by Professor Emeritus Howard Mel, Wendell Latimer's last graduate student and long-time member of the Medical Physics department at Berkeley. He had spent 3 years as Director of the University of California overseas campuses in Southern France. Howard and his wife Nancy are among the five present members who were part of the original group at our first tasting. The group has changed over the years with deaths, divorces, and departures, but many of the present members have been with the group for two decades.

That group consists mostly of couples. Another wine-tasting group is a male group, with wives joining in on special occasions. My entry into this group, The Order of the Purple Tongue, came about in the following way. I mentioned already that Professor Takayuki Fueno of Osaka spent his Sabbatical at Berkeley in 1973. He found housing in a spare room in the house of Noel Engel, at that time an attorney for the IRS. Noel is a wine enthusiast and is now in the business as a wine distributor and consultant. Through meals at the Engel household, Tak also developed a taste for good wine, and he told Noel that he should meet me. Early the following year, after Tak had returned to Japan, I received a call from Noel who introduced himself and invited me to a tasting of the Purple Tongue. I have been a member ever since. The other members are a diverse group that includes an oriental rug merchant, a computer consultant, an engineering professor, a media expert, and a film distributor.

Wine tastings are educational as well as fun. We so often eat and drink without paying much attention to the senses involved. When one does pay attention one becomes aware of changing

Two members of The Order of the Purple Tongue, Emmett Eiland (left) and David Shirley (right), at a wine tasting in Emmett's oriental rug store in 1985. Dave Shirley was the head of the Lawrence Berkeley Laboratory at this time. I have been a member of this wine tasting group for more than 20 years; Emmett was one of the founding members.

aromas and of many nuances of flavor in different parts of the mouth. I also learned that taste is highly subjective, unlike science, which has right and wrong answers. The same wine is occasionally ranked both highest and lowest by different individuals. Nevertheless, there is frequently a consensus in rankings, and these often have no correlation with cost! Professional tasters do not swallow; they could never get through all the many wines they must sample if they did. We, however, only taste seven to eight wines in a tasting, and we do swallow. With a limited number of wines one can enjoy taste sensations without the unpleasant side effects of trying to metabolize too much alcohol.

## Coda

I have had the opportunity and the pleasure to work in several different areas of chemistry. It has certainly added to the fun of my science, but it has also been a challenge to keep learning new areas. One of the most difficult—or more accurately, impossible—jobs has been to try to keep up with the literature of these different areas. The growth of chemistry is illustrated for me dramatically when I compare the thin monthly *JACS* in the 1940s to the present weekly volumes and the few additional journals there were to read then compared to the many today.

Organic chemistry at midcentury was much closer to the organic chemistry of the turn of the century than to the organic chemistry of today. The textbooks of the day were generally classical chemistry with factual data and uses of specific compounds, many name reactions, and virtually no reaction mechanisms. Similarly, organic chemistry in the laboratory when I was a graduate student was not much different from that of earlier decades; the apparatus and techniques were much the same, and there were only a few new reagents, such as lithium aluminum hydride. Many preparations and the apparatus involved have hardly changed at all even now; three-neck flasks, reflux condensers, and addition funnels are still the same and have the same standard taper joints. The greater use of Teflon and such conveniences as magnetic stirrers and syringe pumps mark the principal differences. Work with air-sensitive materials is much more common today and is greatly facilitated by the glove boxes now so common even in organic laboratories.

The big changes are in the use of instruments, notably NMR, first with magnets of increasing field and now with the sophistication of complex pulse techniques. Together with mass spectroscopy and X-ray crystallography, these instrumental methods have made the determination of structure faster and more complete and with requirements of far less material. These new methods are strong and efficient, and they are important. We must all use them, but they are totally different from classical structure determinations and perhaps not as much intellectually fun or satisfying. Chromatographic methods (e.g., GC, thin-layer, HPLC, and column chromatography) have also completely altered the way we work up, purify, and analyze products.

In recent years, a new development has begun to have a profound effect on the way chemistry is done. The disposal of chemicals is now highly regulated and costly. Indeed, disposal of a partially used chemical now frequently costs more than its initial purchase! This disposal problem has already had an impact on some of our preparative experiments.

The developing role of computational methods and theory in organic chemistry has also been impressive. Many of my students over the years have combined theoretical studies with experimental projects. I think this is a marvelous combination and is excellent training. I do not think one should be a "computational organic chemist", that is, an organic chemist who does no experiments at all. I have tried to discourage this direction in my students, but not wholly successfully. I believe that one can be a theoretical chemist (with proper background in mathematics, physics, and physical chemistry) or an organic chemist (who does experiments), but that a purely "theoretical organic chemist" tends to be a master of neither. I like to look at uranocene as an example. This molecule was conceived in theory, and I am certainly proud of its conception, but I am prouder that we could then go into the lab and make it ourselves instead of having to persuade someone else to do it.

The distribution of my students with respect to research topics may be of some interest. By the end of 1990, more than 100 graduate students had worked in my group, 13 in the decade of the 50s and the rest almost evenly divided in subsequent decades.

# A Lifetime of Synergy with Theory and Experiment 259

My research group occasionally organized rafting trips at nearby whitewater rivers. This one took place on the Stanislaus River in Northern California in the summer of 1982. I'm the one with jeans being bounced up on hitting some particularly vigorous rapids. Daughter Susan and her long-time friend Susan Johntz (now Dr. Johntz, a psychiatrist) joined us on this occasion: (from left) our guide (unknown), Susan Streitwieser, Susan Johntz, Matt Lyttle, Dan Bors, me, Scott Gronert, and Mike Kaufman. These outings were invariably lots of fun and popular with the research group. (Reproduced with permission from Rapid Shooters.)

Of these, 82 have now received the Ph.D. degree, and one is currently finishing with me. Almost one-half have done experimental work in the area of carbon acidity and one-third have done theory. The fraction doing theory has increased over this period from rather small in the early decades to almost one-half in some recent years.

Other changes in the practice of chemistry have taken place during my career. When I started at Berkeley, my startup requirements were modest, and normal supplies were provided by the department without cost to me. Modern instruments have made organic chemistry more efficient and, more importantly, have permitted the attack on problems one could not consider in earlier years. Of course, it is also more expensive. As a result, I have spent far more time generating research support than was necessary in earlier decades. This situation is getting worse. Two decades ago, postdoctoral fellowships and especially predoctoral fellowships were more common than today. This potentially disastrous trend for American chemistry means that the aspiring student must seek a mentor who has funds rather than the mentor of choice.

It is instructive to recall the powerful effect that a relatively few fellowships have had on my own career. The AEC postdoctoral fellowship permitted me to work with Professor Roberts and become introduced to MO theory. An NSF Science Faculty Fellowship gave me the opportunity to write the bulk of my MO book in 1959–1960; it would have been impossible at that time without such support. A Guggenheim fellowship in 1969 provided my first real opportunity to spend much time in Europe visiting universities and meeting European colleagues. That trip included a month in Basel, Switzerland, with Professor Edgar Heilbronner and his research group, from whom I learned a great deal about photoelectron spectroscopy; although I never worked directly with this technique, the knowledge gained was important in some of my later research. The subsequent Humboldt and NATO awards were essential for my collaborations with Munich and Erlangen and have led directly to new lanthanide chemistry and computational research. Such fellowship opportunities should be expanded, not contracted. Some personal numbers may be of interest. By the end of 1990 more than 70 postdoctoral students and visiting scientists had worked with me: 4 in the

## A Lifetime of Synergy with Theory and Experiment

*With Professor Duilio Arigoni (left) at the ETH, Zurich, in 1991.*

1950s, 28 in the 1960s, 23 in the 1970s, and 17 in the 1980s. The drop in recent years is entirely a matter of funding. The distribution of these postdoctorates among research areas is approximately the same as for my graduate students.

The increased competitiveness for research funds is discouraging. My chemical history adds up to a reasonable "track record", yet my success rate with research proposals has been no better than 50% during the past decade or two; for comparison, in earlier years I rarely had a proposal turned down. But at least I have been funded and have had the opportunity to do research. I am aware of many colleagues who have lost funding entirely and

With Vladimir Prelog (left) in his office at the ETH, Zurich, in 1991. Since he was forced to retire some years earlier, Professor Prelog had his office officially as a student (see his volume, *My 132 Semesters of Chemistry Studies*, in this series). The picture behind us is a special portrait of Robert Woodward that Prelog keeps in his office.

# A Lifetime of Synergy with Theory and Experiment

*Sue and me in our Munich apartment while on Sabbatical leave in Germany in 1991.*

have stopped research. The future of academic science requires that research directors spend more time thinking about science and less time thinking about financial support. The current emphasis on research proposals and peer review has important negative aspects. Peers often have a tendency to rate more favorably a familiar research area, and thus peer review can discourage work in more arcane or brand-new areas. For such areas and for some agencies it may be better to emphasize track record. There is also an important fiscal aspect. Research funding is often short-range and difficult to plan especially with respect to post-doctorates. Commitments cannot be made while grant proposals are pending, but once granted the interested postdoctorates may well have since accepted their second choice. A better arrangement would undoubtedly be an expanded fellowship program, with each fellowship carrying with it a reasonable amount for supplies and expenses. Research grants could then be reduced in amount. Bookkeeping would also be greatly simplified with such an arrangement. Financial accountability marks another important difference in modern academic research. Running a research group now is much more like a business. Everything from co-worker stipends to shops and the use of instruments, including xeroxing, represents a charge, and I find it necessary to follow these expenditures much more closely than in earlier decades; of course, we now have spreadsheets and computers to help.

I have mentioned already my belief that physical organic chemistry offers great training to students. About one-half of my students have gone to the chemical industry, but there has been little return from industry. The chemical industry does support chemistry departments with fellowships and other funds. But many pharmaceutical houses provide direct support to synthetic organic chemists. Materials research also gets substantial direct support from industry. Physical organic chemistry, however, gets virtually no such support. In my own case, for example, I have received only one small grant from one company early in my career.

One aspect of research groups that has not changed in my four-decade experience has to do with people. Of course, the research director plays a key role in shaping the "personality" of a group, but what may not be immediately obvious is the important role played by some students. Research groups change as new generations of students replace the old, but the character or personality of a group is often determined by one or more especially charismatic, creative, or otherwise individualistic students; "leaders" is a trite but nonetheless applicable term. From my experience as a graduate student and as a professor with many generations of my own Berkeley groups, I have seen how such individuals provide a center for group spirit, chemical discussions, group cohesiveness, and even scientific discipline. They have added greatly to my pleasure, and I am indebted to them. Indeed, the citation of these students calls attention to another fact that emerges clearly in this history: many persons have contributed to whatever success I have had in my chemical career. Teachers, co-workers, and students have had vital roles in my

# A Lifetime of Synergy with Theory and Experiment

For my 65th birthday, John Brauman and Peter Stang organized a dinner in my honor at the House of Prime Rib in San Francisco. About 80 former students and co-workers were there including this group. From left: Kenneth Smith, John Collins, Eusebio Juaristi, Spiro Alexandratos, and Bill Schriver.

scientific development and productivity. Moreover, it is also clear that it is easier to do good research in a place such as Berkeley where one has not only excellent colleagues but also a stimulating atmosphere of creative science. Such an environment also attracts creative and productive students. In a very real sense, my research students have worked *with* me rather than *for* me[421], and they have taught me much. I expect research students to read the literature of their area and to become more expert in the specific area than I am. This increasing familiarity with an area brings ideas, suggestions, and discoveries, and the research problem thereby increasingly becomes the student's and not just mine. This type of collaboration frequently leads to the shared ecstasy of discovery. Even now, there is a unique joy in reviewing puzzling experimental results together and discovering the explanation.

Research institutions such as Berkeley also, of course, require a degree of responsibility to these same colleagues and to the world of scholarship. My hope and expectation is that in my career I have discharged this responsibility to at least the same measure as I have gained from others. My chemistry would also not have been possible without research grants. I am especially indebted to the NSF for continuous support over a long period. The Chemical Dynamics section has been a particularly important part of my research support. The National Institutes of Health provided important support for more than two decades. The Air Force Office of Scientific Research also played a significant role, especially in the era before the Mansfield Amendment eliminated Department of Defense support of any research not directly related to defense activities. The Department of Energy had a major role in supporting my f-element organometallic research from the mid-1970s to 1990. The Petroleum Research Fund had a small but vital role especially in the early years. To all of these agencies, my students, and my colleagues, I offer my heartfelt thanks for giving me the opportunity to have so much fun in the synergy of theory and experiment. My only wish would be to enjoy this life for many more years, or, in the words attributed to Mae West[422], "Too much of a good thing can be wonderful."

Finally, I think my story in chemistry provides the following moral: It is important to have ideas and opinions to help organize one's knowledge of chemistry, but occasionally nature forces one to change one's opinions, and one must be prepared to do so!

# Appendix A: Colleagues and Associates

- Roger Adams, 1899–1971; Professor of Chemistry, University of Illinois. For a biography, see Tarbell, D. S.; Tarbell, A. T. *Roger Adams;* American Chemical Society: Washington, D.C., 1981. (*See* p **20**.)

- Richard Allan Andersen, b. 1942; Professor of Chemistry, University of California, Berkeley. Dick received his B.A. from South Dakota in 1965 and his Ph.D. from Wyoming in 1973. After postdoctorates in Oslo and Imperial College, London, he joined the faculty at Berkeley in 1976. His research is in synthetic inorganic and organometallic chemistry. (*See* p **195**.)

- Richard Frederick William Bader, b. 1931, Professor of Chemistry, McMaster University, Hamilton, Ottawa, Canada, was trained as a physical organic chemist with Gardner Swain at MIT, but became a theoretical chemist during his postdoctoral studies at Cambridge University. He is best known for his work on the topological properties of electron density distributions. (*See* p **157**.)

- Paul Doughty Bartlett, b. 1907; Professor of Chemistry Emeritus, Texas Christian University and Harvard University. *See* Bartlett, P. D. *P.D. and the Bartlett Group at Harvard, 1934–1974;* 1975, a group autobiography on the occasion of the Bartlett Symposium on Physical Organic Chemistry, Fort Worth, TX, August 14–16, 1975. (*See* p **60**.)

- Jerome Abraham Berson, b. 1924; Emeritus Professor of Chemistry, Yale University. Jerry is a graduate of CCNY and received his Columbia Ph.D. in 1949. He did postdoctoral work with Woodward at Harvard, 1949–1950, and taught at USC, 1950–1963, and Wisconsin-Madison, 1963–1969, before moving to Yale. His research has been with reaction mechanisms, usually with strained systems. (*See* p **30**.)

- Frederick G. Bordwell, b. 1916; Professor of Chemistry Emeritus, Northwestern University. (*See* p **100**.)

- Gerald Eyre Kirkwood Branch, 1886–1954; Professor of Chemistry, University of California, Berkeley. Branch graduated from Liverpool in 1911 and received his Ph.D. from Berkeley. He remained on the faculty and did research in physical organic chemistry. (*See* p **45**.)

- John I. Brauman, b. 1937; Professor of Chemistry, Stanford University. John is noted for being among the first to measure acidities in the gas phase. (*See* p **147**.)

- Herbert Charles Brown, b. 1912; R. B. Wetherill Research Professor Emeritus, Purdue University. (*See* p **17**.)

- Melvin Calvin, b. 1911; Professor of Chemistry Emeritus. *See* his profile in this series, *Following the Path of Light*. His book also tells more about early organic chemistry at Berkeley. (*See* p **45**.)

- James Cason, b. 1912; Professor of Chemistry Emeritus, University of California. Jim received his A.B. from Vanderbilt in 1934. After receiving his M.S. at Berkeley in 1935, he obtained his Ph.D. from Yale in 1938. He did postdoctoral work at Harvard, 1938–1940, and taught at DePauw and Vanderbilt before returning to Berkeley in 1945. His research was primarily in the area of fatty acid synthesis, natural products chemistry, and reaction mechanisms. (*See* p **46**.)

- Charles A. Coulson, 1910–1974. Rouse Ball Professor of Mathematics, and later Professor of Theoretical Chemistry, at Oxford University. (*See* p **148**.)

- Donald James Cram, b. 1919; Professor of Chemistry Emeritus, University of California, Los Angeles. *See* his profile in this series, *From Design to Discovery*. (*See* p **58**.)

- William Garfield Dauben, b. 1919; Professor of Chemistry Emeritus, University of California, Berkeley. Bill received his B.A. degree in 1941 from Ohio State and his A.M. and Ph.D. degrees from Harvard. He came to Berkeley in 1945 where his research covers a wide area of synthetic and physical organic, natural products, and photochemistry. (*See* p **46**.)

- Charles Herbert DePuy, b. 1927; Emeritus Professor of Chemistry, University of Colorado, Boulder. Chuck received his B.S. from UC-Berkeley in 1948. He started his Ph.D. work at Columbia and moved with Doering to Yale, receiving a Yale Ph.D. in 1953. After postdoctoral work at UCLA, 1953–1954, he started his academic career at Iowa State before moving to Colorado in 1964. After many years of research in solution physical organic chemistry Chuck started a new career with his study of carbanions in the gas phase. (*See* p **30**.)

- Michael James Steuart Dewar, b. 1918; Professor of Chemistry Emeritus, University of Texas, Austin. *See* his profile in this series, *A Semiempirical Life*. (*See* p **29**.)

- William von Eggers Doering, b. 1917; Professor of Chemistry Emeritus, Harvard University. Doering moved from Columbia to Yale in 1952 and to Harvard in 1968. (*See* p **30**.)

- Gert Ehrlich, b. 1926; Professor, University of Illinois, Urbana, IL. (*See* p **28**.)

- Robert Carlyle Fahey, b. 1936; Professor of Chemistry, University of California, San Diego. (*See* p **64**.)

- Lester Friedman, b. 1928; Professor of Chemistry at Case-Western Reserve University, 1961–1975. (*See* p **5**.)

- Kenichi Fukui, b. 1918. Institute for Fundamental Chemistry, 34–4, Takano-Nishihiraki-cho, Sakyo-ku, Kyoto 606, Japan. (*See* p **80**.)

# A Lifetime of Synergy with Theory and Experiment 269

- E. Peter Geiduschek, b. 1928; Professor of Biology, University of California, San Diego. (*See* p **28**.)

- Frank G. Gresham, 1908–1983; Ph.D. Harvard University, 1936. (*See* p **108**.)

- Ernest Grunwald, b. 1923; Professor of Chemistry Emeritus, Brandeis University. (*See* p **113**.)

- Joel Henry Hildebrand, 1881–1983; Professor of Chemistry Emeritus, University of California, Berkeley. Joel received his B.S. and Ph.D. degrees from Pennsylvania. He returned to the faculty at Penn following a postdoctoral year in Berlin. He came to Berkeley in 1913 and did research on solution chemistry. He retired in 1954 but continued research until his death at the age of 101. Indeed, he was proud to have published more papers after retirement than before. (*See* p **63**.)

- Kieth O. Hodgson, b. 1947; Professor of Chemistry, Stanford University. Kieth is one of the leading experts in XAFS spectroscopy. (*See* p **188**.)

- Roald Hoffmann, b. 1937; Professor of Chemistry, Cornell University. (*See* p **78**.)

- Frederick Richard Jensen, 1925–1987; Professor of Chemistry, University of California, Berkeley. Fritz received B.S. and M.S. degrees from Nevada and in 1955 received the Ph.D. from Purdue. He then came to Berkeley and did research in organic reaction mechanisms, especially of organometallic compounds, and NMR. He retired shortly before his death in 1987. (*See* p **46**.)

- Thomas J. Katz, b. 1936; Professor of Chemistry, Columbia University. (*See* p **113**.)

- Edward Malcolm Kosower, b. 1929; Professor of Chemistry, Tel-Aviv University. Ed graduated from MIT in 1948, took his Ph.D. with Winstein at UCLA in 1952, and did postdoctoral work at Basel and Harvard. After academic positions at Lehigh, Wisconsin-Madison, and SUNY-Stony Brook, he went to Tel Aviv in 1972. His research has covered a broad range of physical organic, biochemical, and biophysical chemistry. (*See* p **5**.)

- Gerd Neustadter LaMar, b. 1937; Professor of Chemistry, University of California, Davis. (*See* p **190**.)

- Wendell Mitchell Latimer, 1893–1955; Professor of Chemistry, University of California, Berkeley; Dean, College of Chemistry, 1941–1949. Latimer graduated from the University of Kansas in 1915. He received his Ph.D. from Berkeley in 1919 and remained on the faculty. His research was mostly in thermodynamics of inorganic compounds. (*See* p **45**.)

- Gilbert Newton Lewis, 1875–1946; Dean of the College of Chemistry, University of California, for most of the period from 1912 to 1941. He received all of his degrees from Harvard and came to Berkeley in 1912. (*See* p **45**.)

- Norman Nahum Lichtin, b. 1922; University Professor Emeritus, Boston University. (*See* p **68**.)

- Jerrold Meinwald, b. 1927; Goldwin Smith Professor of Chemistry, Cornell University. Jerry's research in natural products chemistry has included insect defense mechanism. (*See* p **22**.)

- Herbert Meislich, b. 1920; Professor of Chemistry Emeritus, City University of New York. (*See* p **30**.)

- Ulrich T. Müller-Westerhoff, b. 1937; Professor of Chemistry, University of Connecticut, but for many years a research chemist at IBM, San Jose, CA. His wife, Eda, is an accomplished artist and sculptor. (*See* p **183**.)

- Donald Sterling Noyce, b. 1923; Professor of Chemistry Emeritus, University of California, Berkeley. Don received his A.B. from Grinnell in 1944. After his Ph.D. and a postdoctoral year at Columbia, he came to Berkeley in 1948. Until his retirement in 1986 he did research in kinetics, stereochemistry, and mechanisms of organic reactions. (*See* p **46**.)

- Axel R. Olson, 1889–1954; Professor of Chemistry, University of California. He graduated from Chicago in 1915 and received his Berkeley Ph.D. in 1917. His research was mostly in reaction kinetics and mechanisms and salt effects. (*See* p **46**.)

- G. Edward Pendray, 1901–1987; Founder and Secretary of the American Rocket Society, later incorporated into the American Institute of Aeronautics and Astronomics. He was science editor at the *New York Herald Tribune* for many years and was with Westinghouse, 1936–1945, before forming his own public relations firm. (*See* p **7**.)

- Kenneth Sanborn Pitzer, b. 1914; Professor of Chemistry Emeritus, University of California, Berkeley. Ken received his B.S. at the California Institute of Technology in 1935 and his Ph.D. at Berkeley in 1937. He remained on the faculty at Berkeley but was Director of Research of AEC 1949–1951. He left Berkeley to be President of Rice University, 1961–1968, and Stanford University, 1968–1971, before returning to Berkeley for the remainder of his academic career. His research has been mostly in thermodynamics and the role of relativistic effects in chemistry. (*See* p **65**.)

- John Anthony Pople, b. 1925; for many years Professor of Chemistry at Carnegie-Mellon University, and now Professor at Northwestern University. (*See* p **150**.)

- Henry Rapoport, b. 1918; Professor of Chemistry Emeritus, University of California, Berkeley. Henry received his B.S. and Ph.D. degrees from MIT. He worked in industry and at the U.S. Public Health Service before coming to Berkeley in 1946. Henry's research is in natural products chemistry and synthesis. (*See* p **46**.)

## A Lifetime of Synergy with Theory and Experiment

- Kenneth Norman Raymond, b. 1942; Professor of Chemistry, University of California, Berkeley. Ken received his B.A. from Reed in 1964 and his Ph.D. from Northwestern in 1968 before joining the faculty at Berkeley in 1968. His research has been mostly in bioinorganic chemistry and coordination compounds. (*See* p **185**.)

- John Delano Roberts, b. 1918, Professor of Chemistry, California Institute of Technology. *See* his volume in this series, *The Right Place at the Right Time*. (*See* p **39**.)

- Martin Saunders, b. 1931, Professor of Chemistry, Yale University, overlapped with me at Stuyvesant High School (Class of 1948), then got his B.S. at the City College of New York and his Ph.D. from Harvard with R. B. Woodward. He went to Yale directly from graduate school. His research has been in reaction mechanisms, applications of NMR, molecular mechanics, and most recently with fullerenes encapsulating noble gases. (*See* p **22**.)

- William H. Saunders, Jr., b. 1926; Emeritus Professor of Chemistry, University of Rochester. Bill received his Ph.D. at Northwestern and started his academic career at Rochester following his postdoctoral work with Roberts, 1951–1953. Bill's major independent research has been with elimination reactions and isotope effects. (*See* p **40**.)

- Paul von Rague Schleyer, b. 1930; Professor, Institute of Organic Chemistry, University of Erlangen-Nurnberg. Paul started his academic career at Princeton, his undergraduate alma mater. His volume is scheduled to be in this series. (*See* p **156**.)

- Glenn Theodore Seaborg, b. 1912; University Professor Emeritus, University of California. Glenn received his A.B. from UCLA in 1934 and his Ph.D. from Berkeley in 1937. He stayed on at Berkeley for 2 years before joining the faculty in 1939. He worked on the Manhattan Project, 1942–1946, and was Chairman of AEC, 1961–1971. His research has been mostly on the chemistry of the actinides. (*See* p **195**.)

- Vernon Jack Shiner, Jr., b. 1925; Professor of Chemistry, Indiana University. Jack's research involves principally the use of isotope effects in reaction mechanisms. (*See* p **63**.)

- Peter J. Stang, b. 1941; Professor of Chemistry, University of Utah. Peter was one of the first researchers of vinyl cations derived from vinyl trifluoromethanesulfonates (triflates), which led to many useful syntheses. His most recent work has been with novel iodonium compounds. (*See* p **54**.)

- David Paul Stevenson, b. 1914; for many years a leading research chemist at the Shell Development Company, at that time in Emeryville, California. Dave is now retired and lives in northern California. (*See* p **73**.)

- Thomas Dale Stewart, 1890–1958; Professor of Chemistry, University of California, Berkeley. He received all of his degrees at Berkeley and remained on the faculty. His research was mostly on acid–base equilibria and reaction mechanisms. (*See* p **45**.)

- Robert Wheaton Taft, 1921–1995; Professor of Chemistry, University of California, Irvine. Taft spent a postdoctoral year with Hammett at Columbia and shared a laboratory with me. His earlier academic career was at the Pennsylvania State University. (*See* p **60**.)

- Frank H. Westheimer, b. 1912; Professor of Chemistry Emeritus, Harvard University. Frank's research covers a wide range of important contributions to molecular mechanics, reaction mechanisms, and bio-organic chemistry. (*See* p **57**.)

- Kenneth Berle Wiberg, b. 1927; Professor of Chemistry, Yale University. Ken graduated from MIT in 1948 and completed his Ph.D. at Columbia in 1950. He was on the Faculty of the University of Washington from 1950 to 1960, before moving to Yale. His research has covered a broad range of synthetic chemistry of strained ring compounds and physical organic chemistry including many ab initio calculations. (*See* p **30**.)

- Benjamin Widom, b. 1927; Professor of Chemistry, Cornell University. His research is in statistical mechanics. (*See* p **22**.)

- George Alvin Wiley, 1931–1973, Assistant Professor, 1958–1960. After Berkeley, he moved to the University of Syracuse before resigning to take an active role in the national organization of the Congress Of Racial Equality and helping to create the National Welfare Rights Organization before his tragic death in a boating accident. For a biography, *see* Kotz, N.; Kotz, M. L. *A Passion for Equality;* Norton: New York, 1977. (*See* p **76**.)

- Saul Winstein, 1912–1969, Professor of Chemistry, University of California, Los Angeles. For a biography and bibliography, see *Prog. Phys. Org. Chem.* **1972**, *9*, 1. (*See* p **58**.)

- Alfred Peter Wolf, b. 1923; Department of Chemistry, Brookhaven National Laboratory. Al received an A.B. from Columbia College in 1944 and his Columbia Ph.D. in 1952. His entire subsequent career has been at Brookhaven, where his research has included nuclear chemistry and radiopharmaceuticals. (*See* p **30**.)

- Robert Burns Woodward, 1917–1979, Professor of Chemistry, Harvard University. (*See* p **43**.)

- Richard William Young, b. 1926, has had a full and varied career. He joined the American Cyanamid Company after his Ph.D. and rose to Director of the Agricultural Division. He joined the Polaroid Corporation in 1962 and became Executive Vice President before taking early retirement in 1980.

From 1982 to 1985 he was President of Houghton-Mifflin Publishing Company. From there he was Chairman and CEO of Mentor O&O, a specialized medical products company, until that company merged with Mentor Corporation. He is now a director of several companies. (*See* p **30**.)

- Robert W. Zwanzig, b. 1928; Distinguished Professor of Physical Science Emeritus, University of Maryland; now with NIH. His research has emphasized statistical mechanics. (*See* p **22**.)

# Appendix B: Chemical Genealogy

A number of my students have inquired about our chemical genealogy. Reproduced here is a portion of the extensive genealogy compiled by Vera V. Mainz and Gregory S. Girolami, University of Illinois–Urbana, and reproduced with their permission. The role of "doctor-father" is not always straightforward. An example is the case of Justus von Liebig, who left Erlangen in part because of his involvement in a student riot and learned much of his chemistry from Gay-Lussac in Paris; Liebig's Ph.D. from Erlangen was granted in absentia. Thus, Gay-Lussac is listed as a secondary influence and provides a link to Lavoisier, who collaborated with Bucquet. Berthollet worked jointly with Bucquet and Macquer, both of whom received their doctorates under the direction of Rouelle. In this way, our chemical roots lead to England, Germany, France, and Italy and include some of the great names in the history of chemistry.

| Research Adviser/Teacher | Doctorate | |
|---|---|---|
| | Year | Place |
| Andrew Streitwieser (1927– ) | 1952 | Columbia |
| William v. E. Doering (1917– ) | 1943 | Harvard |
| Reginald P. Linstead (1902–1966) | 1926 | Imperial College |
| George A. R. Kon (1892–1951) | 1922 | Imperial College |
| Jocelyn F. Thorpe (1872–1940) | 1895 | Heidelberg |
| Karl F. v. Auwers (1863–1939) | 1885 | Berlin |
| August W. v. Hofmann (1818–1892) | 1841 | Giessen |
| Justus v. Liebig (1803–1873) | 1822 | Erlangen |
| Joseph L. Gay-Lussac (1778–1850) | 1800[a] | Paris |
| Claude L. Berthollet (1748–1822) | 1778[b] | Paris |
| Jean B. M. Bucquet (1746–1780) | 1770[b] | Paris |
| Pierre J. Macquer (1718–1784) | 1742[b] | Paris |
| Guillaume F. Rouelle (1703–1770) | 1725[c] | Paris |
| Karl F. W. G. Kastner (1783–1857) | 1805 | Jena |
| Johann F. A. Göttling (1753–1809) | 1775[c] | Langensalza |
| Johann C. Wiegleb (1732–1800) | ca 1765[c] | Langensalza |
| Ernst G. Baldinger (1738–1804) | 1760[b] | Jena |
| Christoph A. Mangold (1719–1767) | 1751[b] | Erfurt |
| Georg E. Hamberger (1697–1755) | 1721[b] | Jena |
| Georg W. Wedel (1645–1721) | 1669[b] | Jena |
| Werner Rolfinck (1599–1673) | 1625[b] | Padua |
| Adriaan van den Spieghel (1578–1625) | ca 1603[b] | Padua |

[a]M.A.
[b]M.D.
[c]Apothecary degree.

# References

1. Remick, A. E. *Electronic Interpretations of Organic Chemistry*; Wiley: New York, 1943.
2. Brown, H. C. *Ind. Eng. Chem.* **1944**, *36*, 785.
3. Kharasch, M. S.; Brown, H. C. *J. Am. Chem. Soc.* **1939**, *61*, 2142.
4. Streitwieser, A., Jr. *J. Am. Chem. Soc.* **1944**, *66*, 2127.
5. Sengupta, S. K. *Indian J. Chem.* **1966**, 235.
6. Haworth, R. D.; Moore, B. P.; Pauson, P. L. *J. Chem. Soc.* **1948**, 1045.
7. Dewar, M. J. S. *Nature (London)* **1945**, *156*, 50. *See also* the discussions by Dewer and by T. Nozoe on this subject in their profiles in this series.
8. Doering, W. v. E.; Zeiss, H. H. *J. Am. Chem. Soc.* **1950**, *72*, 147.
9. Doering, W. v. E.; Wolf, A. P. *Perfum. Essent. Oil Rec.* **1951**, *42*, 414.
10. Doering, W. v. E.; Young, R. W. *J. Am. Chem. Soc.* **1950**, *72*, 631.
11. Doering, W. v. E.; Farber, M. *J. Am. Chem. Soc.* **1949**, *71*, 1514.
12. Eliel, E. L. *J. Am. Chem. Soc.* **1949**, *71*, 3970.
13. Alexander, E. R. *J. Am. Chem. Soc.* **1950**, *72*, 3796.
14. Shankland, R. V.; Gomberg, M. *J. Am. Chem. Soc.* **1930**, *52*, 4973.
15. Streitwieser, A., Jr. *J. Am. Chem. Soc.* **1953**, *75*, 5014.
16. Le Roux, L. J.; Sugden, S. *J. Chem. Soc.* **1939**, 1279.
17. Hughes, E. D.; Juliusberger, F.; Masterman, S.; Topley, B.; Weiss, J. *J. Chem. Soc.* **1935**, 1525.
18. For a history of the College of Chemistry to about 1950, *see* Jolly, W. L. *From Retorts to Lasers, the Story of Chemistry at Berkeley*; College of Chemistry, University of California: Berkeley, CA, 1987.
19. Branch, G. E. K.; Calvin, M. *The Theory of Organic Chemistry. An Advanced Course*; Prentice-Hall: New York, 1941.
20. Streitwieser, A., Jr.; Schaeffer, W. D. *J. Am. Chem. Soc.* **1957**, *79*, 6233.
21. Streitwieser, A., Jr.; Andreades, S. *J. Am. Chem. Soc.* **1958**, *80*, 6553.
22. Streitwieser, A., Jr. Thesis, Columbia University, 1952.
23. Streitwieser, A., Jr. *Chem. Rev.* **1956**, *56*, 571.
24. Streitwieser, A., Jr.; Schaeffer, W. D. *J. Am. Chem. Soc.* **1956**, *78*, 5597.
25. Ciereszko, L. S.; Burr, J. G. *J. Am. Chem. Soc.* **1952**, *74*, 145.
26. Burr, J. G.; Ciereszko, L. S. *J. Am. Chem. Soc.* **1952**, *74*, 5431.

27. Bailey, P. S.; Burr, J. G. *J. Am. Chem. Soc.* **1953**, *75*, 2591.
28. Roberts, J. D.; Halmann, M. *J. Am. Chem. Soc.* **1953**, *75*, 5759.
29. Curtin, D. Y.; Crew, M. C. *J. Am. Chem. Soc.* **1954**, *76*, 3719.
30. Roberts, J. D.; Lee, C. C.; Saunders, W. H., Jr. *J. Am. Chem. Soc.* **1954**, *76*, 4501.
31. Fort, A. W.; Roberts, J. D. *J. Am. Chem. Soc.* **1956**, *78*, 584.
32. Cram, D. J.; McCarty, J. E. *J. Am. Chem. Soc.* **1957**, *79*, 2866.
33. Streitwieser, A., Jr.; Schaeffer, W. D. *J. Am. Chem. Soc.* **1957**, *79*, 2888.
34. Streitwieser, A., Jr. *J. Org. Chem.* **1957**, *22*, 861.
35. Brosch, D.; Kirmse, W. *J. Org. Chem.* **1991**, *56*, 907.
36. Ford, G. P. *J. Am. Chem. Soc.* **1986**, *108*, 5104.
37. Streitwieser, A., Jr.; Stevenson, D. P.; Schaeffer, W. D. *J. Am. Chem. Soc.* **1959**, *81*, 1110.
38. Streitwieser, A., Jr.; Schaeffer, W. D.; Andreades, S. *J. Am. Chem. Soc.* **1959**, *81*, 1113.
39. Mislow, K.; O'Brien, R. E.; Schaefer, H. *J. Am. Chem. Soc.* **1962**, *84*, 1940.
40. Streitwieser, A., Jr.; Stang, P. J. *J. Am. Chem. Soc.* **1965**, *87*, 4953.
41. McCauley, D. A.; Lien, A. P. *J. Am. Chem. Soc.* **1953**, *75*, 2411.
42. Brown, H. C.; Smoot, C. R. *J. Am. Chem. Soc.* **1956**, *78*, 2176.
43. Streitwieser, A., Jr.; Reif, L. *J. Am. Chem. Soc.* **1960**, *82*, 5003.
44. Streitwieser, A., Jr.; Downs, W. J. *J. Org. Chem.* **1962**, *27*, 625.
45. Streitwieser, A., Jr.; Reif, L. *J. Am. Chem. Soc.* **1964**, *86*, 1988.
46. Loewus, F. A.; Westheimer, F. H.; Vennesland, B. *J. Am. Chem. Soc.* **1953**, *75*, 5018.
47. Levy, H. R.; Loewus, F. A.; Vennesland, B. *J. Am. Chem. Soc.* **1957**, *79*, 2949.
48. Streitwieser, A., Jr.; Wolfe, J. R.; Schaeffer, W. D. *Tetrahedron* **1959**, *6*, 338.
49. Streitwieser, A., Jr.; Granger, M. R. *J. Org. Chem.* **1967**, *32*, 1528.
50. Streitwieser, A., Jr.; Schwager, I.; Verbit, L.; Rabitz, H. *J. Org. Chem.* **1967**, *32*, 1532.
51. DeWolfe, R. H.; Young, W. G. *Chem. Rev.* **1956**, *56*, 753.
52. Streitwieser, A., Jr. *Solvolytic Displacement Reactions*; McGraw-Hill: New York, 1962.
53. Streitwieser, A., Jr. *J. Am. Chem. Soc.* **1956**, *78*, 4935.
54. Bartlett, P. D. *Nonclassical Ions*; Benjamin: New York, 1965; Brown, H. C.; Schleyer, P. v. R. *The Nonclassical Ion Problem*; Plenum: New York, 1977.
55. Shiner, V. J., Jr. *J. Am. Chem. Soc.* **1953**, *75*, 2925. An earlier publication containing the same idea is that of Lewis, E. S.; Boozer, C. E. *J. Am. Chem. Soc.* **1952**, *74*, 6306.
56. Streitwieser, A., Jr.; Fahey, R. C. *Chem. Ind.* **1957**, 1417.
57. Streitwieser, A., Jr.; Jagow, R. H.; Fahey, R. C.; Suzuki, S. *J. Am. Chem. Soc.* **1958**, *80*, 2326.
58. Kirsch, J. F. In *Isotope Effects in Enzyme Catalyzed Reactions*; Cleland, W. W.; O'Leary, M. H.; Northrup, D. B., Eds.; University Park Press: Baltimore, MD, 1977; p 100.

59. Huskey, W. P.; Schowen, R. L. *J. Am. Chem. Soc.* **1983**, *105*, 5704.
60. Klinman, J. P. In *Enzyme Mechanism from Isotope Effects;* Cook, P. F., Ed.; CRC Press: Boca Raton, FL, 1991; Chapter 4.
61. Boyd, R. J.; Kim, C. K.; Shi, Z.; Weinberg, N.; Wolfe, S. *J. Am. Chem. Soc.* **1993**, *115*, 10147–10152.
62. Zhao, X. G.; Tucker, S. C.; Truhlar, D. G. *J. Am. Chem. Soc* **1991**, *113*, 826–832.
63. Poirier, R. A.; Wang, Y.; Westaway, K. C. *J. Am. Chem. Soc.* **1994**, *116*, 2526–2533.
64. Halevi, E. A. *Prog. Phys. Org. Chem.* **1963**, *1*, 109.
65. Halevi, E. A.; Nussim, M. *Bull. Res. Counc. Isr. Sect. A* **1956**, *5*, 263; Halevi, E. A.; Nussim, M. In *Proceedings of the Sixteenth International Congress of Pure and Applied Chemistry,* Paris, 1957.
66. Halevi, E. A.; Nussim, M.; Ron, A. *J. Chem. Soc.* **1963**, 866.
67. *See also* Ropp, G. A. *J. Am. Chem. Soc.* **1960**, *82*, 4252.
68. Bell, R. P.; Crooks, J. E. *Trans. Faraday Soc.* **1962**, *58*, 1409.
69. Streitwieser, A., Jr.; Klein, H. S. *Chem. Ind.* **1961**, 180.
70. Streitwieser, A., Jr.; Klein, H. S. *J. Am. Chem. Soc.* **1963**, *85*, 2759.
71. Streitwieser, A., Jr.; Klein, H. S. *J. Am. Chem. Soc.* **1964**, *86*, 5170.
72. Kresge, A. J.; Rao, K. N.; Lichtin, N. N. *Chem. Ind.* **1961**, 53.
73. Lichtin, N. N. *Prog. Phys. Org. Chem.* **1963**, *1*, 75.
74. Bernasconi, C.; Koch, W.; Zollinger, H. *Helv. Chim. Acta.* **1963**, *46*, 1184.
75. Streitwieser, A., Jr.; Humphrey, J. S., Jr. *J. Am. Chem. Soc.* **1967**, *89*, 3767.
76. Streitwieser, A., Jr. *Molecular Orbital Theory for Organic Chemists;* Wiley: New York, 1961.
77. Streitwieser, A., Jr.; Roberts, J. D.; Regan, C. M. *J. Am. Chem. Soc.* **1952**, *74*, 4579.
78. Streitwieser, A., Jr.; Roberts, J. D. *J. Am. Chem. Soc.* **1952**, *74*, 4723.
79. Coulson, C. A.; Rushbrooke, G. S. *Proc. Cambridge Phil. Soc.* **1940**, *36*, 193.
80. Coulson, C. A.; Longuet-Higgins, H. C. *Proc. R. Soc.* **1947**, *A192*, 16.
81. Coulson, C. A.; Longuet-Higgins, H. C. *Proc. R. Soc.* **1947**, *A191*, 39.
82. Lichtin, N. N.; Bartlett, P. D. *J. Am. Chem. Soc.* **1951**, *73*, 5530.
83. Lichtin, N. N.; Glazer, H. *J. Am. Chem. Soc.* **1951**, *73*, 5537.
84. Streitwieser, A., Jr. *J. Am. Chem. Soc.* **1952**, *74*, 5288.
85. Hammond, H. A.; Streitwieser, A., Jr. *Anal. Chem.* **1969**, *41*, 2032.
86. Streitwieser, A., Jr.; Hammond, H. A.; Jagow, R. H.; Williams, R. M.; Jesaitis, R. G.; Chang, C. J.; Wolf, R. A. *J. Am. Chem. Soc.* **1970**, *92*, 5141.
87. Streitwieser, A., Jr.; Lewis, A.; Schwager, I.; Fish, R. W.; Labana, S. *J. Am. Chem. Soc.* **1970**, *92*, 6525–6529.
88. *See* Travis, D. *Bull. Am. Acad. Arts Sci.* **1994**, *47*, 23.
89. Roberts, J. D. *Notes on Molecular Orbital Theory;* Benjamin: New York, 1961.
90. Fukui, K.; Yonezawa, T.; Shingu, H. *J. Chem. Phys.* **1952**, *20*, 722.
91. Fukui, K.; Yonezawa, T.; Nagata, C.; Shingu, H. *J. Chem. Phys.* **1954**, *22*, 1433.

92. Dewar, M. J. S.; Dougherty, R. C. *The PMO Theory of Organic Chemistry;* Plenum: New York, 1975.
93. Van-Catledge, F. A. *J. Org. Chem.* **1980,** *45,* 4801.
94. Jorgensen, W. L.; Salem, L. *The Organic Chemist's Book of Orbitals;* Academic: Orlando, FL, 1973.
95. Streitwieser, A., Jr. *Science (Washington, D.C.)* **1981,** *214,* 627.
96. Hall, G. E.; Piccolini, R.; Roberts, J. D. *J. Am. Chem. Soc.* **1955,** *77,* 4540.
97. Summarized in Shatenstein, A. I. *Adv. Phys. Org. Chem.* **1963,** *1,* 156.
98. Streitwieser, A., Jr.; Van Sickle, D. E.; Langworthy, W. C. *J. Am. Chem. Soc.* **1962,** *84,* 244.
99. Streitwieser, A., Jr.; Van Sickle, D. E. *J. Am. Chem. Soc.* **1962,** *84,* 249.
100. Streitwieser, A., Jr.; Langworthy, W. C.; Van Sickle, D. E. *J. Am. Chem. Soc.* **1962,** *84,* 251.
101. Streitwieser, A., Jr.; Van Sickle, D. E. *J. Am. Chem. Soc.* **1962,** *84,* 254.
102. Streitwieser, A., Jr.; Van Sickle, D. E.; Reif, L. *J. Am. Chem. Soc.* **1962,** *84,* 258.
103. Streitwieser, A., Jr.; Koch, H. F. *J. Am. Chem. Soc.* **1964,** *86,* 404.
104. Koch, H. F.; Koch, J. G.; Koch, N. H.; Koch, A. S. *J. Am. Chem. Soc.* **1983,** *105,* 2388.
105. Streitwieser, A., Jr.; Langworthy, W. C. *J. Am. Chem. Soc.* **1963,** *85,* 1757.
106. Streitwieser, A., Jr.; Langworthy, W. C.; Brauman, J. I. *J. Am. Chem. Soc.* **1963,** *85,* 1761.
107. Streitwieser, A., Jr.; Lawler, R. G. *J. Am. Chem. Soc.* **1963,** *85,* 2855; Streitwieser, A., Jr.; Lawler, R. G. *J. Am. Chem. Soc.* **1965,** *87,* 5383.
108. Streitwieser, A., Jr.; Lawler, R. G.; Perrin, C. *J. Am. Chem. Soc.* **1965,** *87,* 5383.
109. Streitwieser, A., Jr.; Ziegler, G. R.; Mowery, P.C.; Lewis, A.; Lawler, R. G. *J. Am. Chem. Soc.* **1968,** *90,* 1357–1358;
110. Streitwieser, A., Jr.; Young, W. R.; Caldwell, R. A. *J. Am. Chem. Soc.* **1969,** *91,* 527.
111. Streitwieser, A., Jr.; Caldwell, R. A.; Young, W. R. *J. Am. Chem. Soc.* **1969,** *91,* 529.
112. Streitwieser, A.; Dixon, R. E.; Williams, P. G.; Eaton, P. E. *J. Am. Chem. Soc.* **1991,** *113,* 357.
113. Streitwieser, A., Jr.; Taylor, D. R. *Chem. Commun.* **1970,** 1248.
114. Maskornick, M. J.; Streitwieser, A., Jr. *Tetrahedron Lett.* **1972,** *17,* 1625.
115. Dixon, R. E.; Streitwieser, A. *J. Org. Chem.* **1992,** *57,* 6125.
116. Streitwieser, A., Jr.; Ni, J. X. *Tetrahedron Lett.* **1985,** *26,* 6317.
117. Dixon, R. E.; Williams, P. G.; Saljoughian, M.; Long, M. A.; Streitwieser, A. *Magn. Res. Chem.* **1991,** *29,* 509.
118. McEwen, W. K. *J. Am. Chem. Soc.* **1936,** *58,* 1124; *see also* still earlier and more qualitative studies by Conant, J. B.; Wheland, G. W. *J. Am. Chem. Soc.* **1932,** *54,* 1212.

119. Hammett, L. P.; Deyrup, A. J. *J. Am. Chem. Soc.* **1932**, *54*, 2721.
120. Hammett, L. P. *Physical Organic Chemistry*, 2nd ed.; McGraw-Hill: New York, 1970.
121. Rochester, C. H. *Acidity Functions*; Academic: Orlando, FL, 1970.
122. Langford, C. H.; Burwell, R. L., Jr. *J. Am. Chem. Soc.* **1960**, *82*, 1503.
123. Rapoport, H.; Smolinsky, G. *J. Am. Chem. Soc.* **1960**, *82*, 934.
124. Kuhn, R.; Fischer, H. *Angew. Chem.* **1961**, *73*, 435.
125. Kuhn, R.; Fischer, H.; Neugebauer, F. A.; Fischer, H. *Liebigs Ann. Chem.* **1962**, *654*, 64.
126. Kuhn, R.; Fisher, H. *Angew Chem. Int. Ed. Engl.* **1964**, *3*, 137.
127. Kuhn, R.; Rewicki, D. *Tetrahedron Lett.* **1964**, 383.
128. Streitwieser, A., Jr.; Brauman, J. I.; Hammons, J. H.; Pudjaatmaka, A. H. *J. Am. Chem. Soc.* **1965**, *87*, 384.
129. Streitwieser, A., Jr.; Ciuffarin, E.; Hammons, J. H. *J. Am. Chem. Soc.* **1967**, *89*, 63.
130. Bowden, K.; Cockerill, A. F. *Chem. Commun.* **1967**, 989–991.
131. Cockerill, A. F.; Lamper, J. E. *J. Chem. Soc.* **1971**, 503–507.
132. Steiner, E. C.; Gilbert, J. M. *J. Am. Chem. Soc.* **1965**, *87*, 382–384.
133. Bowden, K.; Stewart, R. *Tetrahedron* **1965**, *21*, 261.
134. Steiner, E. C.; Starley, J. D. *J. Am. Chem. Soc.* **1967**, *89*, 2751–2752.
135. Bowden, K.; Cockerill, A. F. *J. Chem. Soc.* **1970**, 173–179.
136. Kuhn, R.; Rewicki, D. *Liebigs Ann. Chem.* **1967**, *704*, 9–14.
137. Kuhn, R.; Rewicki, D. *Liebigs Ann. Chem* **1967**, *706*, 250–261.
138. Ritchie, C. D.; Uschold, R. E. *J. Am. Chem. Soc.* **1967**, *89*, 1721, 2752.
139. Ritchie, C. D.; Uschold, R. E. *J. Am. Chem. Soc.* **1968**, *90*, 2821.
140. Ritchie, C. D. *J. Am. Chem. Soc.* **1969**, *91*, 6749.
141. A few leading references are Matthews, W. S.; Bares, J. E.; Bartness, J. E.; Bordwell, F. G.; Comforth, F. J.; Drucker, G. E.; Margolin, Z.; McCallum, R. J.; McCollum, G. J.; Vanier, N. R. *J. Am. Chem. Soc.* **1975**, *97*, 7006; Bordwell, F. G.; Bartness, J. E.; Drucker, G. E.; Margolin, Z.; Matthews, W. S. *J. Am. Chem. Soc.* **1975**, *97*, 3226; Bordwell, F. G.; Matthews, W. S.; Vanier, N. R. *J. Am. Chem. Soc.* **1975**, *97*, 442.
142. For a summary, see Bordwell, F. G. *Pure Appl. Chem.* **1977**, *49*, 963; *Acc. Chem. Res.* **1988**, *12*, 456.
143. Streitwieser, A., Jr.; Chang, C. J.; Hollyhead, W. B. *J. Am. Chem. Soc.* **1972**, *94*, 5292.
144. Streitwieser, A., Jr.; Juaristi, E.; Nebenzahl, L. L. *Comprehensive Carbanion Chemistry*; Elsevier: Amsterdam, Netherlands, 1980; Chapter 7.
145. Streitwieser, A., Jr.; Reuben, D. M. E. *J. Am. Chem. Soc.* **1971**, *93*, 1823.
146. Streitwieser, A., Jr.; Nebenzahl, L. L. *J. Syn. Org. Chem. Jpn.* **1975**, *33*, 889 (in Japanese).
147. Streitwieser, A., Jr.; Granger, M. R.; Mares, F.; Wolf, R. A. *J. Am. Chem. Soc.* **1973**, *95*, 4257.

148. Streitwieser, A., Jr.; Guibé, F. *J. Am. Chem. Soc.* **1978**, *100*, 4532.
149. Cram, D. J. *Fundamentals of Carbanion Chemistry*; Academic: Orlando, FL, 1965.
150. Dorfman, L. M.; Sujdak, R. J.; Bockrath, B. *Acc. Chem. Res.* **1976**, *9*, 352.
151. Streitwieser, A., Jr.; Hollyhead, W.; Pudjaatmaka, A.; Owens, P. H.; Kruger, T.; Rubenstein, P.; MacQuarrie, R.; Brokaw, M.; Chu, W.; Niemeyer, H. M. *J. Am. Chem. Soc.* **1971**, *93*, 5088.
152. Streitwieser, A., Jr.; Kaufman, M. J.; Bors, D. A.; Murdoch, J. R.; MacArthur, C. A.; Murphy, J. T.; Shen, C. C. *J. Am. Chem. Soc.* **1985**, *107*, 6983–6986.
153. Streitwieser, A., Jr.; Hollyhead, W.; Sonnichsen, G.; Pudjaatmaka, A.; Chang, C. J.; Kruger, T. *J. Am. Chem. Soc.* **1971**, *93*, 5096.
154. Kaufman, M. J.; Bors, D. A.; MacArthur, C. A.; Guibé, F.; Murphy, J. T.; Streitwieser, A., unpublished results.
155. Unpublished results.
156. Bordwell, F. G.; Boyle, W. J., Jr.; Hautala, J. A.; Yell, K. C. *J. Am. Chem. Soc.* **1969**, *91*, 4002.
157. Bordwell, F. G.; Boyle, W. J., Jr.; Yee, K. C. *J. Am. Chem. Soc.* **1970**, *92*, 5926.
158. Fukuyama, M.; Flanagan, P. W. K.; Williams, F. T., Jr.; Frainier, L.; Miller, S. A.; Schechter, H. *J. Am. Chem. Soc.* **1970**, *92*, 4689.
159. Kresge, A. J. *J. Am. Chem. Soc.* **1970**, *92*, 3210.
160. Reviewed in Kresge, A. J. *Chem. Soc. Rev.* **1973**, *2*, 475.
161. Roberts, J. D.; Webb, R. L.; McElhill, E. A. *J. Am. Chem. Soc.* **1950**, *72*, 408.
162. Pauling, L. *The Nature of the Chemical Bond*, 2d ed.; Cornell University Press: Ithaca, NY, 1948; p 235. *See also* 3rd ed., 1960; p 314.
163. For an interesting history of research at DuPont, see Hounshell, D. A.; Smith, J. K., Jr. *Science and Corporate Strategy*; Cambridge University Press: Cambridge, England, 1988.
164. From the book title: Littlejohn, D. *The Ultimate Art*; University of California Press: Berkeley, CA, 1992.
165. Campbell, S. F.; Stephens, R.; Tatlow, J. C. *Tetrahedron* **1965**, *21*, 2997.
166. Campbell, S. F.; Leach, J. M.; Stephens, R.; Tatlow, J. C. *J. Fluorine Chem.* **1971**, *1*, 85.
167. Brown, P. J. N.; Stephens, R.; Tatlow, J. C.; Taylor, J. R. *J. Chem. Soc. Perkin Trans.* **1972**, *1*, 937.
168. Streitwieser, A., Jr.; Holtz, D. *J. Am. Chem. Soc.* **1967**, *89*, 692.
169. Andreades, S. *J. Am. Chem. Soc.* **1964**, *84*, 2003.
170. Streitwieser, A., Jr.; Holtz, D.; Ziegler, G. R.; Stoffer, J. O.; Brokaw, M. L.; Guibé, F. *J. Am. Chem. Soc.* **1976**, *98*, 5229.
171. Cram, D. J.; Kingsbury, C. A.; Rickborn, B. *J. Am. Chem. Soc.* **1961**, *83*, 3688.
172. Eigen, M. *Angew. Chem. Int. Ed. Engl.* **1964**, *3*, 1.
173. Lin, A. C.; Chiang, Y.; Dahlberg, D. B.; Kresge, A. J. *J. Am. Chem. Soc.* **1983**, *105*, 5380.
174. Koppel, I. A.; Pihl, V.; Koppel, J.; Anvia, F.; Taft, R. W. *J. Am. Chem. Soc.* **1994**, *116*, 8654–8657.

175. Streitwieser, A., Jr.; Marchand, A. P.; Pudjaatmaka, A. H. *J. Am. Chem. Soc.* **1967**, *89*, 693.
176. Holtz, D. *Chem. Rev.* **1971**, *71*, 1.
177. For a more recent review, see Stock, L. M.; Wasielewski, M. R. *Prog. Phys. Org. Chem.* **1981**, *13*, 253.
178. Apeloig, Y. *J. Chem. Soc., Chem. Commun.* **1981**, 396.
179. Schleyer, P. v. R.; Kos, A. J. *Tetrahedron* **1983**, *39*, 1141.
180. Friedman, D. S.; Francl, M. M.; Allen, L. C. *Tetrahedron* **1985**, *41*, 499.
181. Berke, C. M.; Schriver, G. W.; Grier, D.; Collins, J. B.; Streitwieser, A., Jr. *Tetrahedron* **1981**, *37*, 345.
182. Eng, H., unpublished results.
183. Farnham, W. B. *J. Am. Chem. Soc.* **1985**, *107*, 4565.
184. Bayliff, A. E.; Bryce, M. R.; Chambers, R. D.; Matthews, R. S. *J. Chem. Soc., Chem. Commun.* **1985**, 1018–1019.
185. Hansen, R. L. *J. Org. Chem.* **1965**, *30*, 4322.
186. Streitwieser, A., Jr.; Wilkins, C. L.; Kiehlmann, E. *J. Am. Chem. Soc.* **1968**, *90*, 1598.
187. Dafforn, G. A.; Streitwieser, A., Jr. *Tetrahedron Lett.* **1970**, *36*, 3159.
188. Streitwieser, A., Jr.; Scannon, P. J.; Niemeyer, H. M. *J. Am. Chem. Soc.* **1972**, *94*, 7936.
189. Streitwieser, A., Jr.; Shen, C. C. C. *Tetrahedron Lett.* **1979**, *4*, 327.
190. Stratakis, M.; Wang, P. G.; Streitwieser, A. *J. Org. Chem.* **1996**, *61*, 3145.
191. Streitwieser, A., Jr.; Hudson, J. A.; Mares, F. *J. Am. Chem. Soc.* **1968**, *90*, 648.
192. Streitwieser, A., Jr.; Mares, F. *J. Am. Chem. Soc.* **1968**, *90*, 644.
193. Hogen-Esch, T.; Smid, J. *J. Am. Chem. Soc.* **1965**, *87*, 669.
194. Reviews: Smid, J. *Ions and Ion Pairs in Organic Reactions;* Wiley Interscience: New York, 1972; Vol. 1, p 85; Smid, J. *Angew. Chem. Int. Ed. Engl.* **1972**, *11*, 112; Hogen-Esch, T. E. *Adv. Phys. Org. Chem.* **1977**, *15*, 153.
195. Bors, D. A.; Kaufman, M. J.; Streitwieser, A., Jr. *J. Am. Chem. Soc.* **1985**, *107*, 6975.
196. Gronert, S.; Streitwieser, A., Jr. *J. Am. Chem. Soc.* **1986**, *108*, 7016.
197. Streitwieser, A.; Ciula, J. C.; Krom, J. A.; Thiele, G. *J. Org. Chem.* **1991**, *56*, 1074.
198. Bordwell, F. G. *Acc. Chem. Res.* **1988**, *12*, 456.
199. Petrov, E. S.; Terekhova, M. I.; Shatenshtein, A. I. *Zh. Obshch. Khim.* **1974**, *44*, 1118.
200. Antipin, I. S.; Vedernikov, A. N.; Konovalov, A. I. *Zh. Org. Khim.* **1985**, *21*, 1355.
201. Antipin, I. S.; Vedernikov, A. N.; Konovalov, A. I. *Zh. Org. Khim.* **1989**, *25*, 3.
202. Antipin, I. S.; Gareev, R. F.; Vedernikov, A. N.; Konovalov, A. I. *Zh. Org. Khim.* **1989**, *25*, 1153.
203. Konovalov, A. I.; Antipin, I. S. *Metalloorg. Khim.* **1989**, *2*, 177.
204. Gareyev, R.; Streitwieser, A. *J. Org. Chem.* **1996**, *61*, 1742.

205. Kaufman, M. J.; Gronert, S.; Streitwieser, A. *J. Am. Chem. Soc.* **1988**, *110*, 2829.
206. White, J. J., unpublished results.
207. Xie, L.; Bors, D. A.; Streitwieser, A. *J. Org. Chem.* **1992**, *57*, 4986.
208. Xie, L.; Streitwieser, A. *J. Org. Chem.* **1995**, *60*, 1339.
209. Wang, P.; Bors, D. A.; Speers, P., unpublished results.
210. Kilway, K. V., manuscript in preparation.
211. Jackman, L. M.; Haddon, R. C. *J. Am. Chem. Soc.* **1973**, *95*, 3687.
212. Jackman, L. M.; Szeverenyi, N. M. *J. Am. Chem. Soc.* **1977**, *99*, 4954.
213. Jackman, L. M.; Scamoutzos, L. M.; DeBrosse, C. W. *J. Am. Chem. Soc.* **1987**, *109*, 5355.
214. Wen, J. Q.; Grutzner, J. B. *J. Org. Chem.* **1986**, *51*, 4220.
215. Bauer, V. W.; Seebach, D. *Helv. Chim. Acta* **1984**, *67*, 1972.
216. Kaufman, M. J.; Streitwieser, A., Jr. *J. Am. Chem. Soc.* **1987**, *109*, 6092.
217. Kaufman, M. J.; Gronert, S. V.; Bors, D. A.; Streitwieser, A. *J. Am. Chem. Soc.* **1987**, *109*, 602.
218. Ciula, J. C.; Streitwieser, A. *J. Org. Chem.* **1992**, *57*, 431 (correction, p 6686). These results change somewhat when account is taken of the changing spectra with concentration: Gareyev, R.; Ciula, J. C.; Streitwieser, A. *J. Org. Chem.* **1996**, *61*, 4589.
219. Krom, J. A.; Petty, J. T.; Streitwieser, A. *J. Am. Chem. Soc.* **1993**, *115*, 8024.
220. (a) Seeman, J. I. *Chem. Rev.* **1983**, *83*, 83–134; (b) Seebach, D.; Amstutz, R.; Dunitz, J. D. *Helv. Chim. Acta* **1981**, *64*, 2622.
221. (a) Krom, J. A.; Streitwieser, A. *J. Am. Chem. Soc.* **1992**, *114*, 8747; (b) Streitwieser, A.; Krom, J. A.; Kilway, K. V.; Abbotto, A., in preparation.
222. (a) Stratakis, M.; Abu-Hasanayn, F.; Streitwieser, A. *J. Org. Chem.* **1995**, *60*, 4688; (b) Abu-Hasanayn, F.; Streitwieser, A. *J. Am. Chem. Soc.*, in press.
223. Abbatto, A.; Streitwieser, A. *J. Am. Chem. Soc.* **1995**, *117*, 6358.
224. Stratakis, M.; Streitwieser, A. *J. Org. Chem.* **1993**, *58*, 1989.
225. Streitwieser, A., Jr.; Swanson, J. T. *J. Am. Chem. Soc.* **1983**, *105*, 2502.
226. Lee, B. S., unpublished results.
227. Streitwieser, A., Jr. *Acc. Chem. Res.* **1984**, *10*, 353–357.
228. Sethson, I.; Johnels, D.; Lejon, T.; Edlund, U.; Wind, B.; Rabideau, P. W. *J. Am. Chem. Soc.* **1992**, *114*, 953.
229. Gronert, S.; Streitwieser, A. *J. Am. Chem. Soc.* **1988**, *110*, 4418–4419.
230. Streitwieser, A., Jr.; Nair, P. M. *Tetrahedron* **1959**, *5*, 149.
231. Streitwieser, A., Jr.; Brauman, J. I. *Supplemental Tables of Molecular Orbital Calculations*; Pergamon: London, England, 1965.
232. Coulson, C. A.; Streitwieser, A., Jr. *Dictionary of π-Electron Calculations*; Pergamon: London, England, 1965.
233. Wheland, G. W.; Mann, D. E. *J. Chem. Phys.* **1949**, *17*, 264.
234. Streitwieser, A., Jr. *J. Am. Chem. Soc.* **1960**, *82*, 4123.
235. Streitwieser, A., Jr.; Brauman, J. I.; Bush, J. B. *Tetrahedron* **1963**, *19 (Suppl. 2)*, 379.

236. Pariser, R.; Parr, R. G. *J. Chem. Phys.* **1953**, *21*, 568.
237. Pariser, R.; Parr, R. G. *J. Chem. Phys.* **1956**, *24*, 250, 1112.
238. Pople, J. A. *Trans. Faraday Soc.* **1953**, *49*, 1375.
239. Pople, J. A. *Proc. R. Soc.* **1955**, *A233*, 233; Pople, J. A. *J. Phys. Chem.* **1957**, *61*, 6.
240. Häfelinger, G.; Streitwieser, A., Jr.; Wright, J. S. *Chem. Ber.* **1969**, *73*, 456.
241. Streitwieser, A., Jr.; Mowery, P. C.; Jesaitis, R. G.; Lewis, A. *J. Am. Chem. Soc.* **1970**, *92*, 6529–6533..
242. Streitwieser, A., Jr.; Mowery, P. C.; Jesaitis, R. G.; Wright, J. S.; Owens, P. H.; Reuben, D. M. E. In *Proceedings of the Jerusalem Symposium on Quantum Chemistry and Biochemistry II*; The Israel Academy of Sciences and Humanities: Jerusalem, Israel, 1970; p 160.
243. Pople, J. A.; Santry, D. P.; Segal, G. A. *J. Chem. Phys.* **1965**, *43*, S129.
244. Pople, J. A.; Beveridge, D. L. *Approximate Molecular Orbital Theory*; McGraw-Hill: New York, 1970.
245. Streitwieser, A., Jr.; Jesaitis, R. G. *Sigma Molecular Orbital Theory*; Yale University Press: New Haven, CT, 1970; p 197.
246. Streitwieser, A., Jr.; Jesaitis, R. G. *Theor. Chim. Acta.* **1970**, *17*, 165.
247. Owens, P. H.; Wolf, R. A.; Streitwieser, A., Jr. *Tetrahedron Lett.* **1970**, *38*, 3385.
248. Streitwieser, A., Jr. *Sigma Molecular Orbital Theory*; Sinanoglu, O.; Wiberg, K. B., Eds.; Yale University Press: New Haven, CT, 1970; p 197.
249. Owens, P. H.; Streitwieser, A., Jr. *Tetrahedron* **1971**, *27*, 4471.
250. Streitwieser, A., Jr.; Owens, P. H. *Orbital and Electron Density Diagrams: An Application of Computer Graphics*; Macmillan: New York, 1973.
251. Lathan, W. A.; Curtiss, L. A.; Hehre, W. J.; Lisle, J. B.; Pople, J. A. *Prog. Phys. Org. Chem.* **1974**, *11*, 175.
252. Hehre, W. J.; Radom, L.; Schleyer, P. v. R.; Pople, J. A. *Ab Initio Molecular Orbital Theory*; Wiley: New York, 1986.
253. Williams, J. E., Jr.; Streitwieser, A., Jr. *J. Am. Chem. Soc.* **1975**, *97*, 191.
254. Vorpagel, E. R.; Streitwieser, A., Jr.; Alexandratos, S. D. *J. Am. Chem. Soc.* **1981**, *103*, 3777.
255. Streitwieser, A., Jr.; Williams, J. W.; Alexandratos, S.; McKelvey, J. M. *J. Am. Chem. Soc.* **1976**, *98*, 4778.
256. For a recent review, see Streitwieser, A.; Bachrach, S. M.; Dorigo, A.; Schleyer, P. v. R. In *Lithium Chemistry: A Theoretical and Chemical Overview*; Sapse, A.-M.; Schleyer, P. v. R., Eds.; Wiley-Interscience: New York, 1995; pp 1–43.
257. Alexandratos, S.; Streitwieser, A., Jr.; Schaefer, H. F. *J. Am. Chem. Soc.* **1976**, *98*, 7959.
258. Streitwieser, A., Jr.; Waterman, K. C. *J. Am. Chem. Soc.* **1984**, *106*, 3138.
259. Streitwieser, A., Jr.; Collins, J. B.; McKelvey, J. M.; Grier, D.; Sender, J.; Toczko, A. G. *Proc. Natl. Acad. Sci. U.S.A.* **1979**, *76*, 2499.
260. Collins, J. B.; Streitwieser, A., Jr.; McKelvey, J. M. *Comput. Chem.* **1979**, *3*, 79.

261. Bader, R. F. W. *Acc. Chem. Res.* **1985**, *18*, 9.
262. Bader, R. F. W. *Atoms in Molecules—A Quantum Theory*; Oxford University Press: Oxford, England, 1990.
263. Ritchie, J. P.; Bachrach, S. M. *J. Am. Chem. Soc.* **1987**, *109*, 5909.
264. Agrafiotis, D. K.; Tansy, B.; Streitwieser, A. *J. Comput. Chem.* **1990**, *11*, 1101.
265. Agrafiotis, D. K.; Tansy, B.; Streitwieser, A. *QCPE Bulletin* **1991**, *11*, 13.
266. Grier, D. L. Thesis, University of California, 1981.
267. See the summary by Wolfe, S. *Organic Sulfur Chemistry*; Elsevier: Amersterdam, Netherlands, 1985; Chapter 3.
268. Kost, D.; Klein, J.; Streitwieser, A., Jr.; Schriver, G. W. *Proc. Natl. Acad. Sci. U.S.A.* **1982**, *79*, 3922.
269. A similar structure was calculated by Schleyer, P. v. R.; Kos, A. J. *J. Chem. Soc., Chem Commun.* **1982**, 448.
270. Bachrach, S. M.; Streitwieser, A., Jr. *J. Am. Chem. Soc.* **1984**, *106*, 2283.
271. Streitwieser, A.; Kohler, B. *J. Am. Chem. Soc.* **1988**, *110*, 3769.
272. Bader, R. F. W. *Can. J. Chem.* **1986**, *64*, 1036.
273. Streitwieser, A., Jr.; Vorpagel, E. R. *Coll. Czech. Chem. Comm.* **1988**, *53*, 1961.
274. Grier, D. L.; Streitwieser, A., Jr. *J. Am. Chem. Soc.* **1982**, *104*, 3556.
275. Bors, D. A.; Streitwieser, A., Jr. *J. Am. Chem. Soc.* **1986**, *108*, 1397.
276. Streitwieser, A., Jr.; Rajca, A.; McDowell, R. S.; Glaser, R. *J. Am. Chem. Soc.* **1987**, *109*, 4184.
277. Streitwieser, A., Jr.; McDowell, R. S.; Glaser, R. *J. Comput. Chem.* **1987**, *8*, 788.
278. Rajca, A.; Rice, J. E.; Streitwieser, A., Jr.; Schaefer, H. F. *J. Am. Chem. Soc.* **1987**, *109*, 4189.
279. Rajca, A.; Lee, K. H. *J. Am. Chem. Soc.* **1989**, *111*, 4166.
280. Gronert, S.; Glaser, R.; Streitwieser, A. *J. Am. Chem. Soc.* **1989**, *111*, 3111.
281. Siggel, M. R. F.; Thomas, T. D. *J. Am. Chem. Soc.* **1986**, *108*, 4360.
282. Bachrach, S. M.; Streitwieser, A. *J. Comput. Chem.* **1989**, *111*, 514.
283. (a) Thomas, T. D.; Siggel, M. R. F.; Streitwieser, A., Jr. *J. Mol. Struct. Theochem* **1988**, *165*, 309; (b) Siggel, M. R. F.; Streitwieser, A.; Thomas, T. D. *J. Am. Chem. Soc.* **1988**, *110*, 8022; (c) Wiberg, K. B.; Ochterski, J.; Streitwieser, A. *J. Am. Chem. Soc.*, submitted.
284. Brauman, J. I.; Blair, L. K. *J. Am. Chem. Soc.* **1970**, *92*, 5986.
285. Slee, T.; Larouche, A.; Bader, R. W. F. *J. Phys. Chem.* **1988**, *92*, 6219.
286. Bader, R. F. W.; Larouche, A.; Gatti, C.; Carroll, M. T.; MacDougall, P. J.; Wiberg, K. B. *J. Chem. Phys.* **1987**, *87*, 1142.
287. Wiberg, K. B.; Laidig, K. E. *J. Am. Chem. Soc.* **1987**, *109*, 5935.
288. Wiberg, K. B.; Laidig, K. E. *J. Am. Chem. Soc.* **1988**, *110*, 1872.
289. Thomas, T. D. *Inorg. Chem.* **1988**, *27*, 1695.
290. Wiberg, K. B. *Inorg. Chem.* **1988**, *27*, 3694.
291. Speers, P.; Laidig, K. E.; Streitwieser, A. *J. Am. Chem. Soc.* **1994**, *116*, 9257–9261.

## A Lifetime of Synergy with Theory and Experiment

292. See also Boche, G.; Lohrenz, J. C. W.; Ciolowski, J.; Koch, W. In *Supplement S: The Chemistry of Sulphur-Containing Functional Groups*; Patai, S. R., Ed.; Wiley: New York, 1993; pp 340–362; Wiberg, K. B., Castejon, H. *J. Am Chem. Soc.* **1994**, *116*, 10489.
293. Denbigh, K. G. *Trans. Faraday Soc.* **1940**, *36*, 936.
294. Bachrach, S. M.; Streitwieser, A., Jr. *J. Am. Chem. Soc.* **1986**, *108*, 3946.
295. Kaufman, E.; Schleyer, P. v. R.; Houk, K. N.; Wu, Y.-D. *J. Am. Chem. Soc.* **1985**, *107*, 5560.
296. Kaufman, E.; Schleyer, P. v. R. *J. Comput. Chem.* **1989**, *10*, 437.
297. Kaufman, E.; Sieber, S.; Schleyer, P. v. R. *J. Am. Chem. Soc.* **1989**, *111*, 121.
298. Dixon, R. E.; Streitwieser, A.; Laidig, K. E.; Bader, R. F. W.; Harder, S. *J. Phys. Chem.* **1993**, *97*, 3728.
299. Dixon, R. E. Thesis, University of California, 1990.
300. Neuhaus, A.; Abu-Hasanayn, F.; Brunner, K., unpublished results.
301. Glaser, R.; Streitwieser, A. *J. Am. Chem. Soc.* **1987**, *109*, 1258.
302. Glaser, R.; Streitwieser, A. *Pure Appl. Chem.* **1988**, *60*, 195.
303. Glaser, R.; Streitwieser, A. *J. Am. Chem. Soc.* **1989**, *111*, 8799.
304. Mislow, K. *Introduction to Stereochemistry*; Benjamin: New York, 1966.
305. Duke, A. J.; Bader, R. F. W. *Chem. Phys. Lett.* **1971**, *10*, 631.
306. Shi, Z.; Boyd, R. J. *J. Am. Chem. Soc.* **1989**, *111*, 1575.
307. Shi, Z.; Boyd, R. J. *J. Am. Chem. Soc.* **1990**, *112*, 6789.
308. Shi, Z.; Boyd, R. J. *J. Am. Chem. Soc.* **1991**, *113*, 1072.
309. Harder, S.; Streitwieser, A.; Petty, J. T.; Schleyer, P. v. R. *J. Am. Chem. Soc.* **1995**, *117*, 3253–3259.
310. Acree, S. F. *Am. Chem. J.* **1912**, *48*, 352.
311. Choy, G. S.-C.; Glendening, E. D.; Brown, F., unpublished results; see also Lee, I.; Kim, C. K. *J. Phys. Org. Chem.* **1996**, *8*, 473.
312. Cayzergues, P.; Georgoulis, C.; Mathieu, G. *J. Chim. Phys.* **1987**, *84*, 63.
313. However, see the recent comparison of different definitions of charges of Meister, J.; Schwarz, W. H. E. *J. Phys. Chem.* **1994**, *98*, 8245–8252.
314. Reed, A. E.; Weinhold, F. *J. Am. Chem. Soc.* **1985**, *107*, 1919.
315. Foster, J. P.; Weinhold, F. *J. Am. Chem. Soc.* **1980**, *102*, 7211; Reed, A. E.; Curtis, L. A.; Weinhold, F. *Chem. Rev.* **1988**, *88*, 899.
316. Faust, R.; Glendening, E. D.; Streitwieser, A.; Vollhardt, K. P. C. *J. Am. Chem. Soc.* **1992**, *114*, 8263.
317. Glendening, E. D.; Streitwieser, A. *J. Chem. Phys.* **1994**, *100*, 2900–2909.
318. For example, see Houk, K. N.; Tucker, J. A.; Dorigo, A. E. *Acc. Chem. Res.* **1990**, *23*, 107.
319. Frost, A. A.; Musulin, B. *J. Chem. Phys.* **1953**, *21*, 572.
320. (a) Doering, W. E.; Knox, L. H. *J. Am. Chem. Soc.* **1954**, *76*, 3203; (b) Doering, W. E.; Knox, L. H. *J. Am. Chem. Soc.* **1957**, *79*, 352.
321. Breslow, R. *J. Am. Chem. Soc.* **1957**, *79*, 5318.
322. Breslow, R.; Yuan, C. *J. Am. Chem. Soc.* **1958**, *80*, 5991.
323. Sondheimer, F.; Amiel, Y.; Wolovsky, R. *J. Am. Chem. Soc.* **1956**, *78*, 4178.

324. Sondheimer, F.; Amiel, Y.; Wolovsky, R. *J. Am. Chem. Soc.* **1957**, *79*, 4247, 6263.
325. Sondheimer, F.; Wolovsky, R. *J. Am. Chem. Soc.* **1959**, *81*, 4755.
326. Sondheimer, F.; Wolovsky, R.; Gaoni, Y. *J. Am. Chem. Soc.* **1960**, *82*, 755.
327. Hückel, E. In *International Conference on Physics;* The Physical Society: London, England, 1934; p 9.
328. Hückel, E. *Ein Gelehrtenleben, Ernst und Satire;* Verlag Chemie: Weinheim, Germany, 1975.
329. Streitwieser, A., Jr.; Nebenzahl, L. L. *J. Am. Chem. Soc.* **1976**, *98*, 2188.
330. Thiele, J. *Chem. Ber.* **1901**, *34*, 68.
331. Elofson, R. M. *Anal. Chem.* **1949**, *21*, 917.
332. Katz, T. J. *J. Am. Chem. Soc.* **1960**, *82*, 3784, 3785.
333. Moffitt, W. J. *J. Am. Chem. Soc.* **1954**, *76*, 3386.
334. Vogel, E.; Runzheimer, H. V.; Hogrefe, F.; Baasner, B.; Lex, J. *Angew. Chem.* **1977**, *89*, 909.
335. Streitwieser, A., Jr.; Müller-Westerhoff, U. *J. Am. Chem. Soc.* **1968**, *90*, 7364.
336. Zalkin, A.; Raymond, K. N. *J. Am. Chem. Soc.* **1969**, *91*, 5667.
337. Jones, R. G.; Bindschadler, E.; Blume, D.; Karmas, G.; Martin, G. A., Jr.; Thirtle, J. R.; Gilman, H. *J. Am. Chem. Soc.* **1956**, *78*, 2790.
338. Gilman, H.; Jones, R. G.; Bindschadler, E.; Blume, D.; Karmas, G.; Martin, G. A., Jr.; Nobis, J. R.; Thirtle, J. R.; Yale, H. L.; Yoeman, F. A. *J. Am. Chem. Soc.* **1956**, *78*, 6027.
339. Reynolds, L. T.; Wilkenson, G. *J. Inorg. Nucl. Chem.* **1956**, *2*, 246.
340. Fischer, R. D. *Theor. Chim. Acta* **1963**, *1*, 418.
341. For some recent reviews, *see Organometallics of the f-Elements;* Marks, T. J.; Fragala, I. L., Eds.; Reidel: Dordrecht, Holland, 1985; *Chemistry of the Actinide Elements;* Katz, J. J.; Seaborg, G. T.; Morss, L. R., Eds., Chapman and Hall: London, England, 1986.
342. Streitwieser, A., Jr.; Müller-Westerhoff, U.; Sonnichsen, G.; Mares, F.; Morrell, D. G.; Hodgson, K. O.; Harmon, C. A. *J. Am. Chem. Soc.* **1973**, *95*, 8644.
343. Streitwieser, A., Jr.; Müller-Westerhoff, U.; Mares, F.; Grant, C. B.; Morrell, D. G. *Inorg. Synth.* **1979**, *19*, 148.
344. Streitwieser, A., Jr.; Yoshida, N. *J. Am. Chem. Soc.* **1969**, *91*, 7528.
345. Karraker, D. G.; Stone, J. A.; Jones, E. R., Jr.; Edelstein, N. *J. Am. Chem. Soc.* **1970**, *92*, 4841.
346. Starks, D. F.; Parsons, T. C.; Streitwieser, A., Jr.; Edelstein, N. *Inorg. Chem.* **1974**, *13*, 1307.
347. *See also* Goffart, F.; Fuger, J.; Brown, D.; Dayckaerts, G. *Inorg. Nucl. Chem. Lett.* **1974**, *10*, 413.
348. Streitwieser, A., Jr.; Dempf, D.; LaMar, G. N.; Karraker, D. G.; Edelstein, N. *J. Am. Chem. Soc.* **1971**, *93*, 7343.
349. Streitwieser, A., Jr.; Harmon, C. A. *Inorg. Chem.* **1973**, *12*, 1102.
350. Streitwieser, A., Jr.; Walker, R. *J. Organomet. Chem.* **1975**, *97*, C41.

## A Lifetime of Synergy with Theory and Experiment

351. Harmon, C. A.; Bauer, D. P.; Berryhill, S. R.; Hagiwara, K.; Streitwieser, A., Jr. *Inorg. Chem.* **1977**, *16*, 2143.
352. Streitwieser, A., Jr.; Burghard, H. P. G.; Morrell, D. G.; Luke, W. D. *Inorg. Chem.* **1980**, *19*, 1863.
353. For some reviews, *see* Streitwieser, A., Jr. *Topics in Nonbenzenoid Aromatic Chemistry*; Hirokawa: Tokyo, Japan, 1973; Vol. 1, p 221; Streitwieser, A., Jr. *Nachr. Chem. Techn.* **1976**, *24*, 313; Streitwieser, A., Jr. In *Organometallics of the f-Elements*; Marks, T. J.; Fischer, R. D., Eds.; Reidel: Dordrecht, Holland, 1979; pp 149–177; Streitwieser, A., Jr. *Inorg. Chim. Acta* **1984**, *94*, 171–177.
354. Harmon, C. A.; Streitwieser, A., Jr. *J. Am. Chem. Soc.* **1972**, *94*, 8926.
355. Grant, C. B.; Streitwieser, A., Jr. *J. Am. Chem. Soc.* **1978**, *100*, 2433.
356. Edelstein, N.; Streitwieser, A., Jr.; Morrell, D. G.; Walker, R. *Inorg. Chem.* **1976**, *15*, 1397.
357. Luke, W. D.; Streitwieser, A., Jr. In *Lanthanide and Actinide Chemistry and Spectroscopy*; Edelstein, N. M., Ed.; ACS Symposium Series 131; American Chemical Society: Washington, DC, 1980; p 93.
358. Luke, W. D.; Streitwieser, A., Jr. *J. Am. Chem. Soc.* **1981**, *103*, 3241.
359. Moore, R. M., Jr.; Streitwieser, A., Jr.; Wang, H.-K. *Organometallics* **1986**, *5*, 1418.
360. Häfelinger, G.; Regelmann, C. *J. Comput. Chem.* **1985**, *6*, 368.
361. Starks, D. F.; Streitwieser, A., Jr. *J. Am. Chem. Soc.* **1973**, *95*, 3423.
362. Solar, J. P.; Burghard, H. P. G.; Banks, R. H.; Streitwieser, A., Jr. *Inorg. Chem.* **1980**, *19*, 2186.
363. Eisenberg, D. C.; Streitwieser, A.; Kot, W. K. *Inorg. Chem.* **1990**, *29*, 10.
364. LeVanda, C.; Streitwieser, A., Jr. *Inorg. Chem.* **1981**, *20*, 656.
365. LeVanda, C.; Solar, J. P.; Streitwieser, A., Jr. *J. Am. Chem. Soc.* **1980**, *102*, 2128.
366. Zalkin, A.; Templeton, D. H.; LeVanda, C.; Streitwieser, A., Jr. *Inorg. Chem.* **1980**, *19*, 3560.
367. Boussie, T. R.; Moore, R. M., Jr.; Streitwieser, A.; Zalkin, A.; Brennen, J.; Smith, K. A. *Organometallics* **1990**, *9*, 2010.
368. Boussie, T. R.; Streitwieser, A. *J. Org. Chem.* **1993**, *58*, 2377.
369. Streitwieser, A.; Barros, M. T.; Wang, H. K.; Boussie, T. *Organometallics* **1993**, *12*, 5023–5024.
370. Mares, F.; Hodgson, K. O.; Streitwieser, A., Jr. *J. Organomet. Chem.* **1970**, *C68*, 24.
371. Hodgson, K. O.; Mares, F.; Starks, D. F.; Streitwieser, A., Jr. *J. Am. Chem. Soc.* **1973**, *95*, 8650.
372. Hodgson, K. O.; Raymond, K. N. *Inorg. Chem.* **1972**, *11*, 171, 3030.
373. Kinsley, S. A.; Streitwieser, A., Jr.; Zalkin, A. *Organometallics* **1985**, *4*, 52.
374. Eisenberg, D. C.; Kinsley, S. A.; Streitwieser, A. *J. Am. Chem. Soc.* **1989**, *111*, 5769.
375. Boussie, T. R.; Eisenberg, D. C.; Rigsbee, J. T.; Streitwieser, A.; Zalkin, A. *Organometallics* **1991**, *10*, 1922.

376. Clark, J. P.; Green, J. C. *J. Organomet. Chem.* **1976**, *112*, C14.
377. Clark, J. P.; Green, J. C. *J. Chem. Soc. Dalton Trans.* **1977**, 505.
378. Fragala, I. In *Organometallics of the f-Elements*; Reidel: Dordrecht, Holland, 1979; p 421.
379. Green, J. C.; Payne, M. P.; Streitwieser, A., Jr. *Organometallics* **1983**, *2*, 1707.
380. Chang, A. H. H.; Pitzer, R. M. *J. Am. Chem. Soc.* **1989**, *111*, 2500.
381. Greco, A.; Cesca, S.; Bertolini, G. *J. Organomet. Chem.* **1976**, *113*, 321.
382. Rösch, N.; Streitwieser, A., Jr. *J. Organomet. Chem.* **1978**, *145*, 195.
383. Rösch, N.; Streitwieser, A., Jr. *J. Am. Chem. Soc.* **1983**, *105*, 7237.
384. Rösch, N. *Inorg. Chim. Acta* **1984**, *94*, 297.
385. Streitwieser, A., Jr.; Kinsley, S. A.; Rigsbee, J. T.; Fragala, I.; Ciliberto, E.; Rösch, N. *J. Am. Chem. Soc.* **1985**, *107*, 7786.
386. Jensen, C. Thesis, University of California, 1991.
387. His first book of poetry is Hoffmann, R. *The Metamict State*; University of Central Florida Press: Orlando, FL, 1987.
388. Hoffmann, R.; Torrence, V. *Chemistry Imagined*; Smithsonian Institution Press: Washington, DC, 1993.
389. Kilimann, U.; Herbst-Irmer, R.; Stalke, D.; Edelmann, F. T. *Angew. Chem. Int. Ed. Engl.* **1994**, *33*, 1618.
390. Fischer, R. D. *Angew. Chem. Int. Ed. Engl.* **1994**, *33*, 2165.
391. Dolg, M.; Fulde, P.; Küchle, W.; Neumann, C.-S.; Stoll, H. *J. Chem. Phys.* **1991**, *94*, 3011.
392. Dolg, M.; Fulde, P.; Stoll, H.; Preuss, H.; Chang, A.; Pitzer, R. M. *Chem. Phys.* **1995**, *195*, 71–82.
393. Neumann, C.-S.; Fulde, P. *Z. Phys.* **1989**, *B74*, 277.
394. Moore, R. M., Jr. Thesis, University of California, 1985.
395. See for example the tables of Shannon, R. D.; Prewitt, C. T. *Acta Cryst.* **1969**, *B25*, 925; *Acta Cryst.* **1970**, *B26*, 1046.
396. Streitwieser, A.; Smith, K. A. *J. Mol. Struct. (Theochem.)* **1988**, *163*, 259.
397. Yoshida, Z. *Topics Curr. Chem.* **1973**, *40*, 47.
398. Eliel, E. R. *From Cologne to Chapel Hill*; American Chemical Society: Washington, DC, 1990; p 68.
399. Smith, K. A.; Streitwieser, A., Jr. *J. Org. Chem.* **1983**, *48*, 2629.
400. Smith, K. A.; Waterman, K. C.; Streitwieser, A., Jr. *J. Org. Chem.* **1985**, *50*, 3360.
401. Speer, D. V.; DiMagno, S. G., Unpublished results.
402. Waterman, K. C.; Streitwieser, A., Jr. *J. Am. Chem. Soc.* **1984**, *106*, 3874.
403. Waterman, K. C.; Speer, D. V.; Streitwieser, A.; Look, G. C.; Stack, J. G.; Nguyen, K. O. *J. Org. Chem.* **1988**, *53*, 583–588.
404. Feng, A. S.; Speer, D. V.; DiMagno, S. G.; Konings, M. S.; Streitwieser, A. *J. Org. Chem* **1992**, *57*, 2902–2909.
405. Koch, A. S.; Waterman, K. C.; Banks, K.; Streitwieser, A. *J. Org. Chem.* **1990**, *55*, 6166.
406. Jung, M. E.; Buszek, K. R. *J. Org. Chem.* **1985**, *50*, 5440.

407. Koch, A. S.; Feng, A. S.; Hopkins, T. A.; Streitwieser, A. *J. Org. Chem.* **1993**, *58*, 1409.
408. Kosower, E. M.; Ramsey, B. G. *J. Am. Chem. Soc.* **1958**, *81*, 856.
409. DiMagno, S. G.; Waterman, K. C.; Speer, D. V.; Streitwieser, A. *J. Am. Chem. Soc.* **1991**, *113*, 4679.
410. Vajda, E.; Tremmel, J.; Rozsondai, B.; Hargittai, E.; Maltsev, A. K.; Kagramanov, N. D.; Nefedov, O. M. *J. Am. Chem. Soc.* **1986**, *108*, 4352.
411. Speer, D. V. Thesis, University of California, 1989.
412. DiMagno, S. G.; Streitwieser, A., in preparation.
413. Carter, P. W.; DiMagno, S. G.; Porter, J. D.; Streitwieser, A. *J. Phys. Chem.* **1993**, *97*, 1085.
414. Weiss, R.; Pomrehn, B.; Hampel, F.; Bauer, W. *Angew. Chem. Int. Ed. Engl.* **1995**, *34*, 1319.
415. Streitwieser, A., Jr.; Ward, H. R. *J. Am. Chem. Soc.* **1962**, *84*, 1065; Streitwieser, A., Jr.; Ward, H. R. *J. Am. Chem. Soc.* **1963**, *85*, 538.
416. Rodemeyer, S. A. Thesis, University of California, 1965.
417. Bittman, R. Thesis, University of California, 1965.
418. *J. Org. Chem.* **1990**, *55*, 7A.
419. Snow, C. P. *The Affair*; Scribner's: New York, 1960. (Reprinted in *Strangers and Brothers* ; 1972.; Vol. 2.
420. For an early history, *see* Cochrane, R. C. *National Academy of Sciences, the First Hundred Years, 1863–1963*; National Academy of Sciences: Washington, DC, 1978.
421. Alexandratos, S. *Chem. Eng. News* **1994**, *72*, 5.
422. Weintraub, J. *The Wit and Wisdom of Mae West*; Putnam: New York, 1967.

# Index

**A**

Ab initio MO calculations, computer methods, 149–153
Abelson, Phillip, editor of *Science*, 78
Acidity
  conversion of kinetic to equilibrium, 171
  deuterium isotope effects, 67
  different media, 128*t*–129*t*
Actinide chemistry group, Lawrence Berkeley Laboratory, 195
Actinide sandwich compounds, d and f orbitals, 208–209
Activation energies, solvolysis reactions, 52
Adams, Roger, meeting at Westinghouse Science Talent Search, 20
AEC, *See* Atomic Energy Commission
Aggregation numbers, determination in dilute solution, 134–135
Aggregation states
  alkylation and aldol addition reactions, 137
  lithium and cesium salts, 136–137
Air Force Office of Scientific Research
  research grants, 266
  support of molecular orbital theory work, 91–92

Alexandratos, Spiro
  birthday dinner (photo), 265
  theoretical and experimental chemistry, 153
Alkylation, stereochemical study, 54
Alkyldiazonium cation, branch point for alternative reactions, 51
All-valence-electron semiempirical methods, computer calculations, 149
Allinger, Norman (Lou), editorial board meeting in 1994 (photo), 246
Allylic systems, transition state bonding, 175
Allylides
  production, 227
  reversible one-electron oxidations, 234
American Chemical Society, denial of student membership, 26–27
Ames, Bruce, biochemistry professor, 10
Andersen, Richard A.
  actinide chemistry group, 195
  visit to *Boca Verita* (Mouth of Truth) (photo), 196
Andreades, Sam
  acetolysis reaction in presence of dibutyl ether, 49
  cyclooctatetraene derivatives, 182
  tritium exchange, 118

Andrews, Roy Chapman, dinosaur bone, 8
Anharmonicity, bond vibrations, 66–67
Anionic hyperconjugation
 acidities, 123
 β-fluoroethyl anion, 108f
 importance, 123–124
[8]Annulene compounds
 hydrolysis products, 218t
 metallocenes, 189
Antidepressant drugs, opposite effect, 112
Arigoni, Duilio
 at the ETH, Zurich, in 1991 (photo), 261
 speaker at Natick Conference in 1969 (photo), 187
Army service, Medical Corps, 24–27
Aromatic chemistry, home laboratory, 17
Aromatic hydrocarbons, series of p$K$ values, 99
Aromatic ring protons, kinetic acidity studies, 94
Arylmethanes
 Brønsted correlation, 103f
 isotope exchange reactions, 104–105
Astronomy
 first scientific love, 7
 Junior Astronomy Club, 7–8
Atomic Energy Commission postdoctoral fellowship, MIT, 39–44
Awards
 ACS California Section Award, 112, 248
 ACS Petroleum Award, 248
 Bavarian Academy of Sciences corresponding member, 251
 Berkeley Citation, 251
 Cope Scholar Award, 152
 Elected to National Academy of Sciences, 249
 Humboldt Award for Senior U.S. Scientists, 202–204, 209, 249–250
 Norrish Award in Physical Organic Chemistry, 250

B

Bachrach, Steve
 integrated populations of oxygen in compounds, 167
 research group in 1985 (photo), 229
Bader, Richard W. F.
 giving a talk in 1990 (photo), 158
 programs for VAX computer system, 168
 spatial electron populations, 157
Barros, Maria Teresa, preparation of bridged uranocene, 197
Bartlett, Paul D.
 organic faculty luncheon meeting, March 1983 (photo), 48
 seminars at Harvard, 60
 Senior Scientist Awards in 1976 (photo), 209
Basis functions, 1s, 2s, and 2p atomic orbitals, 162–164
Bending force constants, deuterium isotope effect, 64
Bergman, Bob, Christmas skit in 1985 (photo), 242
Berkeley Camera Club, membership, 251
Berryhill, Stuart, conformational studies, 192
Berson, Bella, picnic excursion to Bear Mountain (photo), 38
Berson, Jerome A.
 Doering research group, 30
 in 1949 (photo), 34
 philosophically one of us, 32
 picnic excursion to Bear Mountain (photo), 38
Bittman, Bob
 microwave discharge products, 237
 Streitwieser's birthday party (photo), 119
Bixler, Mark, Streitwieser's birthday party (photo), 119
Bond weakening, secondary isotope effect, 63
Bonding MO, overlapping orbitals, 211f, 212
Bordwell, Fred, p$K_a$ values in DMSO, 100

# A Lifetime of Synergy with Theory and Experiment

Bors, Daniel A.
  cesium and lithium salts in THF, 127
  rafting trip in 1982 (photo), 259
  sulfonyl SO bond in sulfones and derived carbanions, 165
Boussie, Tom, new routes to substituted cyclooctatetraenes, 197
Brain chemistry, malfunction, 111
Branch, Gerald
  physical organic chemistry course, 176–177
  textbook on physical organic chemistry, 45–46
Brauman, John
  FORTRAN programming, 147
  party at the Koch's in 1960 (photo), 91
  playing guitar at group party in 1961 (photo), 147
Braun, Ingrid, Professor Hofacker's secretary (photo), 212
Breslow, Ronald, Science Talent Search, 22
Bridged uranocene, preparation, 197–199
Bridgehead compounds, equilibrium acidities, 121
Brønsted correlations, localized and delocalized carbanions, 102–107
Brønsted family, determination, 105–106
Brønsted slopes, exchange transition state, 106
Brown, Herbert C., paper on chlorinations with sulfuryl chloride, 17
*Burg Streitwiesen*
  legend, 11
  visit in 1980 (photo), 14
Buttoncraft Company, odd jobs, 9

## C

Caldwell, Richard A., hydrogen isotope exchange reactions, 95
California, terminal leave junket, 27
Calvin, Melvin
  organic faculty luncheon meeting, March 1983 (photo), 48
  photosynthesis, 46

Calvin, Melvin—*Continued*
  physical organic chemistry course, 176–177
  textbook on physical organic chemistry, 45–46
Carbon acidity, 85–143
Carbon–lithium bond, covalent character, 154–156
Carbanion(s)
  Brønsted correlations, 102–107
  tetrahydrofuran, 125–143
Carbanion chemistry, measures of relative acidities of hydrocarbons, 85
Carbonium ion
  activation energies, 53
  deuterium isotope effects, 63–68
  displacement on primary centers, 54–55
  historical context, 51 (footnote)
  Hückel $\pi$-energy differences, 71$f$
  quantitative values of substituent effects, 61
  reaction sequence, 51
Carboxylate resonance, 166–170
Carboxylic acid, deprotonation effect on electron population, 167
Cason, James
  at old chemistry building, January 1961 (photo), 47
  synthetic chemistry, 46
Cerocene
  chemistry, 201–216
  oxidizing agents, 213–214
  proposed models and calculations, 216
  $\chi\alpha$MO calculations, 209–212
Cesium, volatility, 94
Cesium and lithium ion-pair acidities, sulfones and sulfoxide, 133$t$
Cesium enolate of 1,3-diphenylacetone, aggregation plot, 135$f$
  linear aggregation plot, 136$f$
Cesium ion pair, enolate concentration, 139$f$
Cesium ion-pair acidities
  benzyl sulfides and selenide in THF, 131$t$
  dithianes in THF, 132$t$
  salts, 127–143

Chemical genealogy, 275–276
Chemical industry, research
  support, 264
Chemical shifts, effect of structure, 192
Chemistry, fun for children, 8
Chemistry institutes, old buildings in
  the heart of Munich, 206
Chemistry laboratory, changes,
  257–258
Chinese food, initiation into research
  group, 33
Ciula, Jim, aggregation plots, 135
Cohen, Ted, Conference on Carbanion
  Chemistry, Ottawa, July 1989 (photo),
  142
Colleagues and associates, short
  biographies, 267–273
Collins, John
  birthday dinner (photo), 265
  electron density function, 157
Collum, Dave, Conference on
  Carbanion Chemistry, Ottawa, July
  1989 (photo), 142
Columbia College
  first year, 23–24
  heavy schedule, 27
Columbia University
  graduate school, 29–39
  students paid the bills, 38–39
Compton, Karl T., MIT in 1946
  (photo), 15
Computational methods, changes, 258
Computational quantum chemistry
  perception of qualitative concepts,
    169–170
  potential applications, 146
Computer calculations, MO theory,
  145–149
Computer graphics, orbitals and
  electron density functions, 150–151
Conductivity measurements, lithium
  and cesium salts, 130
Conjugated hydrocarbons, acidities
  in different solvents, 100
Contact ion pairs
  carbanion chemistry, 126
  diagrams, 127f
Controversies, journal manuscripts,
  245–246

Coulson, Charles
  development and application of
    HMO theory (photo), 148
  HMO computations, 148
Covalent bonding, carbon–lithium
  bond, 154–156
Cram, Donald, phenonium ion
  rearrangements, 58
Cycloalkanes, tritium exchange rates
  with CsCHA, 96f
Cyclohexylamine
  catalyst reactivity, 94–95
  equilibrium acidities, 97–101
  kinetic acidities, 85–97
  relative ion-pair p$K$ values of cesium
    salts, 101f
Cyclooctatetraene
  first preparation of calcium
    derivative, 201
  rapid electron transfer between
    derivatives, 201
  relative hydrolytic structure
    stabilities, 217–218
  reversible reduction, 182
Cyclooctatetraene derivatives
  ring–metal covalency, 202
  sandwich structure, 200
Cyclopentadienyl ligand, follow-up
  chemistry, 188
Cyclopropenyl cations, stability, 224

D

d orbitals
  ring–metal bonding of cyclooctate-
    traene derivatives of actinides, 202
  stabilization of carbanions by
    adjacent sulfur, 152–153
  $\chi\alpha$MO calculations, 208–209
Dauben, William
  high-pressure apparatus, 232
  organic faculty luncheon meeting,
    March 1983 (photo), 48
  solvolysis chemistry, 58
  surprise birthday party, about 1970
    (photo), 59
  synthetic chemistry, 46
Davis, Watson, Science Talent Search,
  21–22

Decker, O. M., fifth-grade teacher, 8
Delocalization, ionic transition state, 175–176
Densely charged compounds, 223–236
Department of Energy, research grants, 266
DePuy, Charles
  Conference on Carbanion Chemistry, Ottawa, July 1989 (photo), 142
  Doering research group, 30
Deuterium isotope effect
  applications, 66
  secondary, 63–68
  solvolysis, 64–68
Dewar, Michael J. S.
  electrophilic substitution, 71–72
  HOMO and LUMO, 75
  paper on stipitatic acid, 29
  physical and theoretical organic chemist (photo), 72
Dicarbanion, coulomb interactions for point charge model of dication salt, 143f
DiMagno, Steve
  allylides soluble in chloroform, 235
  NMR and X-ray crystallography, 225
1,1'-Dimethylcerocene, X-ray crystal structure, 215f
Dimroth, Karl, arranged interview with Professor Hückel, 182
Dipole stabilization, ion pair charge–charge interaction, 127
Disposal of chemicals, changes, 258
Dissociation constants, free ions, 131
Doering, William von Eggers
  $4n + 2$ rule, 181
  at Yale (photo), 31
  intellectual analytical view of chemistry, 42
  research group, 30–39
Doering research group, about 1947–1948 (photo), 38
Double-bond–no-bond resonance, β-fluoroethyl anion, 108f
DuPont Company
  1939–1940 New York World's Fair, 108
  consulting, 107–109

E

Eclipse, nature award winner at Berkeley Camera Club (photo), 253
Edelstein, Norman
  magnetic phenomena and spectroscopy, 189
  Second International Conference on Lanthanides and Actinides, Lisbon, April 1987 (photo), 199
Edgerton, Harold, talent search lecturer, 20
Editorship, impact on literature of organic chemistry, 243–246
Education
  AEC postdoctoral fellowship, 39–44
  Columbia College, 23–24, 27–28
  elementary school, 8
  graduate school, 29–39
  high school, 5
Ehrlich, Gert, physical chemistry laboratory, 28
Eiland, Emmett, Order of the Purple Tongue, 256
Eisenberg, Dave
  cyclooctatetraene derivative of uranium, 201
  research group in 1985 (photo), 229
Electrocyclic ring closure, indolizine, 225
Electron density
  difference plot, 160, 161f–163f
  irregularity, 178
Electron density function
  applications, 154–166
  HCLi plane of methyllithium, 155f
Electron transfer, high- to low-energy orbitals, 210, 211f
Electrons close to nucleus, relativistic correction, 208
Electrophilic substitution, polycyclic aromatic hydrocarbons, 70–71
Eliel, Ernest, speaker at Natick Conference in 1969 (photo), 187
Eminent Chemists, ACS video series, 78
Energy level diagram, Hückel $4n + 2$ rule, 181

Engel, Noel, Order of the Purple Tongue, 256
English, journal manuscripts, 244
Equilibrium acidities, electrostatic considerations, 122
Errors, journal manuscripts, 245
Ethics, scientific, 246–248
Exchange rates, fluorene with methanolic sodium methoxide, 120
Experimental chemistry, chemical understanding, 2
Experimental data, ethics, 247–248

F

f orbitals
  organometallic chemistry, 181–221
  ring–metal bonding of cyclooctatetraene derivatives of actinides, 201–202
  $\chi\alpha$MO calculations, 208–209
Faculty Club, weekly organic faculty luncheon, 48
Fahey, Bob, $\alpha$-deuterium isotope effect, 64
Farber, Milton
  bicyclic compounds, 37
  picnic excursion to Bear Mountain (photo), 38
Fellowships
  career effect, 260
  research funding, 263–264
$\beta$-Fenchol, acid-catalyzed dehydration, 35–36
Feng, Amy
  ensured presence at commencement, 251
  heterocyclic nitrogen in imidazole, 230
Ferrocene, ring–metal bonding, 182
Fishing
  every summer, 115
  fly-fishing, 251
  fly-fishing in Oregon, 166–167
Fitzsimmons General Hospital, U.S. Army Medical Corps, 25–26
Fluorene, chlorination, 17–20
Fluorinated bicyclics, equilibrium acidity, 122

Fluorinated compounds, research, 124–125
Fluorocarbon chemistry, DuPont discussion group, 108
$\beta$-Fluoroethyl anion, 107, 108$f$
Food, San Francisco area, 255
FORTRAN, HMO program, 147
Fox, Marye Anne, Conference on Carbanion Chemistry, Ottawa, July 1989 (photo), 142
Fraud, ethics, 246–247
Free samples, chemicals, 17
Friedman, Lester
  home laboratory, 7
  Organic Specialties, 5–7
Fueno, Takayuki
  Order of the Purple Tongue, 256
  visit to Napa Valley, 81
Fukui, Kenichi
  frontier orbitals, 75
  international visits, 80–81
  visit to Napa Valley winery in 1972 (photo), 79
Fun aspect of chemistry, 92–94
Future trends, 176–180

G

Gasteiger, Johnny
  at dinner in 1995 (photo), 205
  *Nachsitzung* following seminar at University of Munich in 1991 (photo), 207
Gasteiger, Ullie, at dinner in 1995 (photo), 205
Geiduschek, E. Peter, physical chemistry laboratory, 28
German lessons, Humboldt Foundation research, 105
G.I. Bill, benefits, 24
Glaser, Rainer
  lithium and sodium salts of oximes and enol ethers, 172
  polar bonds to silicon, 165
  uranocene chemistry and theoretical calculations on electron density functions, 171

Glendening, Eric, natural energy decomposition analysis, 178
Gloria, Lynne
  mounted and displayed chemistry set, 119
  secretary (photo), 241
  work on textbook, 240
Glovebox–spectrometer combination, single-indicator technique, 133
Goldberg, Gloria, introduced second wife, 113
Gong, Leyi, research group in 1985 (photo), 229
Goodrich, Frank, good friend, 112
Graduate school, Columbia University, 29–39
Green, Jenny, at lunch in Napa Valley in 1981 (photo), 190
Gresham, Frank, DuPont discussion group, 108
Grier, David
  polarizations of sigma electrons in carbonyl groups, 165
  projection function approach, 159–160
Grob, Cyril, research in Basel, 15
Gronert, Scott
  cesium and lithium salts in THF, 127
  polar bonds to silicon, 165
  rafting trip in 1982 (photo), 259
  research group in 1985 (photo), 229
Gubbian Lock, prize-winning slide, Berkeley Camera Club (photo), 254

H

H–D asymmetry
  partially asymmetric reductions, 57–58
  stereochemistry of alkylation, 54
Half-sandwich structures, uranocene and thorocene, 197
Hammett acidity function method, weak acids and bases, 99
Heathcock, Clayton
  Christmas skit in 1985 (photo), 242
  fly-fishing in summer of 1970 (photo), 239
  *JOC* editor-in-chief, 243
Heathcock, Clayton—*Continued*
  synthetic applications of aldol addition reactions, 134
  textbook coauthor, 238–243
Heller, Adam, microwave discharge products, 237
Heterocycle polycations, substituted with pyridium and related cation groups, 223–236
Hildebrand, Joel, good ideas, 63–64
Hobbies
  fly-fishing, 251
  photography, 251–252
Hodgson, Kieth, crystal structures, 188
Hofacker, Ludwig
  *Lehrstuhl für Theoretische Chemie*, 206
  luncheon meeting of *Lehrstuhl für Theoretische Chemie* (photo), 210
Hoffmann, Roald
  *Chemistry Imagined*, 215–216
  extended Hückel theory, 78
  speaker at Natick Conference in 1969 (photo), 187
  theoretical studies of organometallic compounds, 206
  video interview, 78–80
Holtz, David
  anionic hyperconjugation, 123
  rate constants for tritium exchange, 117–118
  Streitwieser's birthday party (photo), 119
Home laboratory
  organic chemistry, 5–9
  parents' support, 13
Honeymoon, with children, 115
House, Herbert, Science Talent Search, 22
Hückel, Erich, worked between two worlds, 181–182
Hückel $4n + 2$ rule, energy level diagram, 181
Hückel molecular orbital calculations, enjoyment, 145
Hückel molecular orbital theory
  carbanion systems, 85
  interest in graduate school, 68
  ionization equilibria, 68–69

Hückel molecular orbital theory—
*Continued*
  nodal properties of MOs, 77
  role of qualitative perturbation
    approaches, 75
  suggested parameter values for
    heteroatoms, 77t
Huisgen, Rolf, *Nachsitzung* following
  seminar at University of Munich
  in 1991 (photo), 207
Humboldt Foundation, history,
  202–204
Hunig, Siegfried, speaker at Natick
  Conference in 1969 (photo), 187
Hydrocarbons
  isotope exchange reactions, 104
  proton-exchange reaction rates, 96
Hydrogen isotope exchange reactions,
  rates, 94–97
Hydrolysis rate, electron-attracting
  substituents, 216
Hyperconjugation, chemical importance,
  123–124

I

Independent learning, *Electronic
  Interpretations of Organic Chemistry*,
  16–17
Indicator acidity scales, cesium and
  lithium salts in THF, 127
Individual contributions, chemistry, 245
Indolizines, stacking aggregation, 236
Industrial chemistry, consulting at
  DuPont, 107–109
Instruments, changes, 258
Internal return
  carbanion reaction mechanism, 120f
  proton-transfer reactions, 121–122
Intuitive jump, research idea, 39
Inverse isotope effect, C–H bond, 64
Inverse sandwich compounds,
  definition, 223
Ion pair(s), electronic spectra, 127
Ion-pair acidities
  cesium and lithium salts in THF,
    127–143
  dianions, 141
  triple-ion structure, 143

Ion-pair equilibria
  acidity differences between com-
    pounds, 99
  equilibrium constants, 97
Ion-pair process, stabilization by
  electron-donating groups, 176
Ion-pair reactions
  protodedeuteration of toluene-$\alpha$-$d$
    with LiCHA, 87
  transition states, 170–176
Ionic bonding, carbon–lithium bond,
  154–156
Ionic model, computational study of
  cyclopentadienyllithium, 156–157
Ionic $pK_a$ values, compounds in
  cyclohexylamine, 100
Ionic radius
  f-element organometallic compounds
    of cyclooctatetraene, 218, 219t
  mutual repulsion among lone pairs,
    219–220
Ionic separation, solvent dipoles, 126
Ionization free energies and Hückel
  delocalization energy differences,
  correlation curve, 70f
Island Park, Idaho, summer cabins, 115
Isotope effect, protodedeuteration of
  toluene-$\alpha$-$d$ with LiCHA, 87
Isotopic label, deuterium, 63–68
IUPAC rules, games, 5

J

Jagow, Robert H., computer calcula-
  tions, 146
Jensen, Frederick Richard
  organic faculty luncheon meeting,
    March 1983 (photo), 48
  reaction mechanisms and physical
    organic chemistry, 46–47
Johnson, Bill, dinner at Fisher Island,
  Miami, in 1989 (photo), 75
Johnson, Keith, $\chi\alpha$ scattered wave
  method, 208
Johnston, Hal, notification from
  National Academy of Sciences, 249
Johntz, Susan, rafting trip in 1982
  (photo), 259

# A Lifetime of Synergy with Theory and Experiment 301

*Journal of Organic Chemistry*, 243–246
Juaristi, Euseabio, birthday dinner (photo), 265

## K

Katz, Tom, temporary replacement teacher, 113
Kaufman, Michael J.
 cesium and lithium salts in THF, 127
 rafting trip in 1982 (photo), 259
 research group in 1985 (photo), 229
Kende, Andrew, Science Talent Search, 22
Kinsley, Steve
 oxidation of Ce(III) compounds, 213
 ytterbium sandwich structure derivative, 200
Klein, Harvey, secondary deuterium isotope effects on acidity, 67
Klinghoffer, Leon, hijacking of *Achille Lauro*, 56–57
Koch, Andy
 chemistry research, 90
 research group in 1985 (photo), 229
Koch, Heinz
 family careers, 87, 89
 in Grimentz, Switzerland, in 1972 (photo), 88
 KISPOC at Fukuyoku University in 1993, (photo), 89
 party at the Koch's in 1960 (photo), 91
 toluene substituent effect, 87
Koch, Judy
 chemistry research, 90
 Ithaca College, 87
Koch, Nanci, chemistry research, 90
Kohler, Boris, projection function differences, 164
Kosower, Edward M.
 home laboratory, 7
 laboratory notebook, 13–15
 MIT in 1946 (photo), 15
 Organic Specialties, 5–7
 Stuyvesant High School, around 1944 (photo), 6
 textbook coauthor, 241

Kosower, Edward M.—*Continued*
 Westinghouse Science Talent Search project, 21
Krom, Jim, UV–visible spectra showing changes in equilibrium distribution, 135–136

## L

Laboratory notebook, 1943 and 1944, 13–15
LaMar, Gerd, paramagnetic NMR, 190–191
Langworthy, Bill
 Berkeley lab in August 1961 (photo), 90
 exchange reactions of $\alpha$-deuterated polycyclic methylarenes, 90–91
 party at the Koch's in 1960 (photo), 91
Lanthanide(s), cyclooctatetraene derivatives, 200
Lanthanide contraction, effective ionic radius, 218–219
Latimer, Wendell, chemical engineering and organic chemistry, 45
Lawler, Ronald, acidities of aromatic ring protons, 94
Lawrence Berkeley Laboratory
 Materials and Molecular Research Division, 195
 Nuclear Division, 195
LeVanda, Carole, preparation of thorocene, 196
Levitz, Mortimer, picnic excursion to Bear Mountain (photo), 38
Lewis, Adolphus, electrophilic substitution, 71
Lewis structures, accurate accounting for electrons, 177
Liberal arts education, Columbia College, 23–24
Lichtin, Norman, ionization equilibria, 68
Lithium enolates
 reaction mechanism, 137
 spectral analysis, 139–140
Lithium salt
 carbanion from acetaldehyde oxime, lowest energy structure, 173*f*

Lithium salt—*Continued*
  ion-pair acidities, 127–143
  *p*-phenylisobutyrophenone, aggregation plot, 140*f*
Lone pair, effect in MO theories, 161
Long, Louis, Jr., speaker at Natick Conference in 1969 (photo), 187
Luke, Wayne, conformational studies, 192
Lyttle, Matt, rafting trip in 1982 (photo), 259

## M

Machine language, program for HMO-type calculations, 146
Magnetic moment, unpaired electrons on uranium, 191–192
Maier, Wilhelm
  at lunch in Napa Valley in 1981 (photo), 190
  computer system, 166
Mares, Frank, organoactinide chemistry, 188
Massachusetts Institute of Techology, AEC postdoctoral fellowship, 39–44
McDermitt, Todd, synthetic applications of aldol addition reactions, 134
McDowell, Bob, highly polar bonds with little multiple-bond character, 165
Medical technician's training, Army, 25–26
Meinwald, Jerrold, Stuyvesant High School, 22
Meislich, Herb, Doering research group, 30
Mel, Howard, wine tastings at famous chateaux, 255
Metallation, electrophiles, 126
Metallocenes, HOMO–LUMO interactions, 183*f*
Methyl tosylate, kinetic measurements of reactions, 138–140
Methylarene-$\alpha$-*d*, cyclohexylamide exchange rates with lithium, 92*f*
Methyllithium, projection function, 158–159
Microscope set, fascination, 8

Microwave discharge products, organic compounds, 237
Mislow, Kurt, symposium in honor of Schleyer's 60th birthday (photo), 156
Moffitt, William, ferrocene ring–metal bonding, 182
Molecular mechanics, conformations of large molecules, 180
Molecular orbital theory
  computer calculations, 145–149
  solvolysis review, 61–62
  suitability for textbooks, 241–243
*Molecular Orbital Theory for Organic Chemists*, first book, 68–78
Moore, Bob
  research group in 1985 (photo), 229
  substituted aryluranocenes, 216
Morrell, Dennis, magnetic susceptibility of uranocene derivatives, 189
Müller-Westerhoff, Ullie
  research position at IBM, 188
  uranocene synthesis, 183–184
Mulliken populations, basis function, 161–164
Muxfeld, Max, speaker at Natick Conference in 1969 (photo), 187

## N

Nair, P. Madhavan, IBM 701 computations, 146
National Academy of Sciences, history, 249
National Institutes of Health, research grants, 266
National Science Foundation, research grants, 266
Natural bond orbitals, wave function analysis, 178
Natural population analyses, electron calculation, 178
Neptunocene, charge transfer, 202
New York Public Library, Russian journal reference, 16
Ni, Jin-Xiang
  research group in 1985 (photo), 229
  tritium exchange in aqueous sodium hydroxide, 104

Nodal properties of molecular orbitals, Hückel theory, 77
Nonclassical carbonium ion, research efforts, 62
Norbornyl cation, controversy, 62
Normal organic acids, definition, 121
Notebook, research ideas, 28
Noyce, Donald
 at old chemistry building, January 1961 (photo), 47
 organic faculty luncheon meeting, March 1983 (photo), 48
 physical organic chemistry course, 176–177
 reaction mechanisms and physical organic chemistry, 46–47
 research group meetings, 60
 synthetic chemistry, 46
Nucleophilic solvent addends, additional intermediates, 50

O

Olson, Axel, displacement reactions and their stereochemistry, 46
Opera
 common interest, 114
 performances, 252, 255
Optical activity, deuteriobutanol, 41–42
Optically active deuterium compounds, application to reaction mechanisms, 53
Orbital occupancies, simple computation, 178
Orbital symmetry perturbation, applications to organic problem, 76
Order of the Purple Tongue, wine-tasting group, 255–256
Organic chemistry
 changes, 257–266
 support, 62
 syntheses, 29
Organic plasma chemistry, 237–238
Organic Specialties
 business enterprise, 5–7
 chlorination of aniline in "outdoor facilities", 15–16

Organic Specialties—*Continued*
 2-chlorofluorene, 18
 free samples, 17
 Russian journal reference, 16
Organoactinide compounds, structures and hydrolysis, 216–221
Organofluorine carbanions, 115–125
Organofluorine chemistry, consulting at DuPont, 107–109
Organolanthanide chemistry, 200–201
Organolithium compounds, bonding, 154–156
Owens, Peter
 computational quantum chemistry, 150
 computer graphics, 150–151
Oxonium ion, dioxane–methanol solvolysis, 49
Ozawa, Shuji
 cyclooctatetraene derivatives, 182
 near Mt. Fuji in 1971 (photo), 81
 tour of Japan, 81

P

$\pi$-carbon, effective electronegativity, 148
$\pi$-electronic systems, applications to chemistry, 72
Paramagnetism, uranocene, 189
Peer review, financial support, 261–263
Pendray, G. Edward
 astronomy club meeting, 7–8
 Science Talent Search, 22
Pericyclic reactions, Woodward–Hoffmann rules, 77
Perpyridinium allylides, pK values, 231
"Peter Grivich," Golden Gate Bridge, 121
Petroleum Research Fund, research grants, 266
Pfeiffer, Heinrich
 25th anniversary of reestablishment of Humboldt Foundation, 1978 (photo), 208
 met in Germany, 205–206
Phenyl group rotation, barrier, 194
Phenylacetylene, ion-pair equilibria, 101

Physical organic chemistry,
   support, 62
Pitzer, Kenneth, statistical thermodynamics, 65
Plagiarism, ethics, 247
Plumbing, remodeling, 10
Polar substituent constants, solvolytic
   displacement reactions, 60–61
Polarization
   distribution of electron popula-
      tion, 167
   sigma bonds, 169
   sigma electrons, 164–165
   structures, 168f
   sulfur stabilization mechanism, 131
Polyarylmethanes, hydrogen isotope
   exchange rates, 102–107
Pople, John
   at dinner in 1995 (photo), 151
   GAUSSIAN series of programs,
      151–152
   STO-NG basis sets, 150
Porter, Jack, polymer and electrochemical studies, 236
Potassium salts, structures, 219t
Prelog, Vladimir, at the ETH, Zurich, in
   1991 (photo), 262
Primary carbon, stereochemistry,
   48–58
Primary isotope effects, definition, 63
Projection function
   difference plot, 161f–163f
   electron density, 157
   isoelectronic compounds, 160
   methyllithium, 158–159
Proton transfer reactions, highly bent
   transition states, 171
Publications
   *Acidity of Hydrocarbons* series, 94
   approach to writing, 73–74
   effect of computerization, 166
   effect of wife's death, 113
   first paper, 17–20
   HMO computations in book form, 147
   *Introduction to Organic Chemistry*,
      238–243
   *Molecular Orbital Theory for Organic
      Chemists*, 68–78
Purpurogallin, structure, 29

Pyridine
   less nucleophilic, 228
   microwave discharge products, 237
Pyridinium ring, acidity of cyclopentadiene, 232–233
Pyridiniumcarbons
   allylides, 227, 229f, 230f
   development, 224–227
   pentapyridinium cyclopentadienyl
      allylide, 232
   X-ray structure, 228f

Q

Quantum organic chemistry, ab initio,
   149–153
Quinine, Ed Kosower's proposed
   synthesis, 20

R

Rabitz, Herschel, Streitwieser's
   birthday party (photo), 119
Racemization
   proposed mechanism, 52–53
   rate and mechanism, 55
   solvolysis studies, 49
   totally chiral pathways, 173
Radioactive research, rules, regulations,
   and liabilities, 196
Rajca, Andrzej, electronic structure of
   threefold symmetric species, 165
Rapoport, Henry, synthetic chemistry, 46
Rats, organic chemistry laboratories, 47–48
Raymond, Ken
   actinide chemistry group, 195
   coordination compounds of
      actinides, 186
   determination of redox potentials, 213
   tenth anniversary of discovery of
      uranocene (photo), 185
   uranocene crystal structure
      determination, 185–186
Rearrangement products, carbonium
   ion intermediates, 54
Reif, Fred, family history, 55–57

Reif-Lehrer, Liane
  disproportionation of ethylbenzene, 54–55
  family history, 55–57
  Harvard Medical School, 57
  in laboratory about 1959 (photo), 56
Relativistic effects, χαMO calculations, 208–209
Renton, Paul, Army assignment, 26
Research groups, student leaders, 264
Research reports, importance, 43
Research support
  changes, 260
  competitiveness, 261–264
Research topics, changes, 258–260
Resonance structures
  allylic systems, 169
  chemical understanding, 176
  displacements on allylic fluorides, 175f
  energy, 168f
  precise use, 177–178
Retirement, projects, 2
Reviewers, journal manuscripts, 244
Rigsbee, John, research group in 1985 (photo), 229
Ring deuteration, reaction rates, 67
Ring–metal bonding, ferrocene, 182
Ring–metal distances, electrostatic argument, 219
Ring protonation, hydrolysis rate-determining step, 217–218
Ring protons, structure effect on chemical shifts, 192
Ring rotation, barriers, 192–194
Roberts, Edith, dinner at Fisher Island, Miami, in 1989 (photo), 75
Roberts, John D. (Jack)
  dinner at Fisher Island, Miami, in 1989 (photo), 75
  in 1975 (photo), 40
  research laboratories, 39
Rodemeyer, Steve, microwave discharge products, 237
Roosevelt, Eleanor, meeting at Westinghouse Science Talent Search, 20

Rösch, Notker
  collaboration on χαMO calculations of uranocene and thorocene, 206, 208
  research program on lanthanide chemistry, 214–215
  Second International Conference on Lanthanides and Actinides, Lisbon, April 1987 (photo), 199
  theoretical chemist at Technical University (photo), 213

S

Sabbatical year, Berkeley 1959–1960, 72–74
San Francisco Bay Area, amenities, 252, 255
Sandwich compound of cerium(IV), reported preparation, 202
Sandwich structure
  complex salts, 218
  cyclooctatetraene derivatives of lanthanides, 200
  1,1'-dimethylcerocene, 214, 215f
Sasse, Phil, research group in 1985 (photo), 229
Saunders, Martin, Stuyvesant High School, 22
Saunders, William H., Jr., in 1953 (photo), 41
Schaad, Lawrence, Science Talent Search, 22
Schaefer, Henry F. (Fritz), at dinner in 1995 (photo), 151
Schaeffer, William D.
  optically active butanol, 50
  stereochemistry of primary carbon, 48–49
Schilling, Birgitte, sigma polarization in cations, 164
Schleyer, Paul
  C–Li bond, 156, 164
  Conference on Carbanion Chemistry, Ottawa, July 1989 (photo), 142
  editorial board meeting in 1994 (photo), 246
  symposium in honor of 60th birthday (photo), 156

Schriver, Bill, birthday dinner (photo), 265
Scientific fraud, ethics, 246–247
Secondary isotope effects, definition, 63
Seeman, Jeff, ACS meeting in Anaheim, March 1995 (photo), 3
Self-consistent field $\pi$-methods, computer calculations, 149
Sewer cleaning, odd jobs, 10
Shapley, Harlow, talent search lecturer, 20
Sharpshooters' medal, Army, 25
Shiner, Jack, hyperconjugation, 63
Shirley, David, Order of the Purple Tongue, 256
Shriner, Ralph, editor of *Chemical Reviews*, 60
Siberia, separation from the rest of Doering's laboratories, 32
Siggel, Michele
  high acidity implied in electronic structure, 166
  high negative charge on oxygens, 167
  photoelectron spectroscopy and VAX calculations, 167
Single-indicator technique, absorbance, 133
Singular value decomposition, UV–visible spectra, 136
Slide of the Year, Northern California Council of Camera Clubs, 252
Smith, Kennith
  birthday dinner (photo), 265
  inverse sandwich compounds, 223
  pyridiniumcarbons, 224–227
  unexpected chemistry, 235–236
$S_N2$ displacement reactions
  average relative rates of alkyl systems, 61$t$
  review of solvolysis, 58–62
$S_N2$ transition states
  ab initio calculations, 66
  calculations, 173–174
Solvation of ions, problem for computer calculations, 170
Solvent-separated ion pairs
  carbanion chemistry, 126
  diagrams, 127$f$

Solvolysis
  activation energies, 52
  polycyclic arylmethyl tosylates in acetic acid, 69–70
  racemization, 49
Solvolysis chemistry, graduate work, 34–35
Solvolytic displacement reactions, review of the subject, 58–60
Sonnichsen, George
  early chemical studies of uranocene, 188
  postdoctoral researcher, 121
Spectroscopic data, compound characterization, 43–44
Speer, Drew
  NMR and X-ray crystallography, 225
  research group in 1985 (photo), 229
Spin polarization
  direction of chemical shift, 191
  substituted uranocene, 191$f$
Stang, Peter
  carbonium ion intermediate, 54
  symposium in honor of Schleyer's 60th birthday (photo), 156
Starks, David, plutonium analog of uranocene, 195
Stephens, R., highly fluorinated bicyclo[2.2.1]heptanes, 117
Stereochemistry
  computational studies of isomerizations and racemizations, 172
  optically active primary alcohol, 39–42
  primary carbon, 48–58
Stevenson, Dave, course on MO theory, 73
Stewart, Dale, hydration of alkenes in sulfuric acid, 45
Streitwieser, derivation and pronunciation, 10–11
Streitwieser, Andrew, Jr.
  1967 ACS Award in Petroleum Chemistry (photo), 248
  25th anniversary of reestablishment of Humboldt Foundation, 1978 (photo), 208
  38th birthday present, Gilbert chemistry set (photo), 119

Streitwieser, Andrew, Jr.—*Continued*
  ACS meeting in Anaheim, March 1995 (photo), 3
  "Andyland" T-shirt (photo), 179
  at old chemistry building, January 1961 (photo), 47
  at the ETH, Zurich, in 1991 (photo), 261, 262
  awards dinner in Las Vegas in 1982 (photo), 28
  birth, 9
  California Section Award, 1964 (photo), 111
  Christmas skit in 1985 (photo), 242
  Conference on Carbanion Chemistry, Ottawa, July 1989 (photo), 142
  discussing chemistry in Cologne in 1991 (photo), 184
  fly-fishing in summer of 1970 (photo), 239
  home laboratory, 7
  in apartment of Jack and Edith Roberts in 1952 (photo), 110
  in office in 1984 (photo), 220
  late 1927 or early 1928 (photo), 10
  Latimer Hall laboratory in 1963 (photo), 86
  Munich beer (photo), 203
  near Mt. Fuji in 1971 (photo), 81
  organic faculty luncheon meeting, March 1983 (photo), 48
  Organic Specialties, 5–7
  playing ball at group picnic in 1960s (photo), 93
  presenting Roger Adams Award (photo), 80
  rafting trip in 1982 (photo), 259
  Raymond and Beverly Sackler Distinguished Lecturer in Chemistry (photo), 250
  recent picture (photo), 24
  sabbatical leave in Germany in 1991 (photo), 263
  school picture at age of 6 (photo), 11
  Second International Conference on Lanthanides and Actinides, Lisbon, April 1987 (photo), 199
  seminar on carbon acidity research in 1963 (photo), 98

Streitwieser, Andrew, Jr.—*Continued*
  Senior Scientist Awards in 1976 (photo), 209
  sheepskin coat for sabbatical in Germany in 1976 (photo), 206
  South America tour to see solar eclipse (photo), 252
  speaker at Natick Conference in 1969 (photo), 187
  Stuyvesant High School, around 1944 (photo), 6
  symposium in honor of Schleyer's 60th birthday (photo), 156
  temporary office in *Lehrstuhl für Theoretische Chemie* (photo), 211
  trip to Japan (photo), 82
  U.S. Army Medical Corps (photo), 25–26
  visit to Germany, 1930 (photo), 11
  visit to Napa Valley winery in 1972 (photo), 79
  walking on campus with first baby, David (photo), 46
  Westinghouse Science Talent Search project (photo), 21
  with roadster in Buffalo, 1930 (photo), 11
Streitwieser, Andrew (father)
  about 1946 (photo), 12
  carpenter born in Munich, 9
  death, 249
  debilitating stroke, 249
  late 1927 or early 1928 (photo), 10
  varied employment, 9
Streitwieser, David (son)
  at Berkeley home in 1977 (photo), 116
  Berkeley Day Care Center, 109
  emergency room physician, 115
  physician, 10
  six years old in 1960 (photo), 74
  work on textbook, 240
Streitwieser, Mary Ann (wife)
  at California Section Award, 1964 (photo), 111
  bouts of depression, 109–111
  in apartment of Jack and Edith Roberts in 1952 (photo), 110
  suicide, 112

Streitwieser, Mary Ann (wife)—
*Continued*
  walking on campus with first baby, David (photo), 46
Streitwieser, Sophie (mother)
  about 1946 (photo), 12
  family background, 13
  from Black Forest, 9
  in 1988, at age of 84 (photo), 13
  visit to Germany, 1930 (photo), 11
Streitwieser, Sue (wife)
  25th anniversary of reestablishment of Humboldt Foundation, 1978 (photo), 208
  at lunch in Napa Valley in 1981 (photo), 190
  family background, 114
  in 1966, before engagement (photo), 114
  *Nachsitzung* following seminar at University of Munich in 1991 (photo), 207
  near Mt. Fuji in 1971 (photo), 81
  sabbatical leave in Germany in 1991 (photo), 263
  sheepskin coat for sabbatical in Germany, in 1976 (photo), 206
  trip to Japan (photo), 82
  work on textbook, 240
Streitwieser, Susan (daughter)
  at Berkeley home in 1977 (photo), 116
  foster home, 109
  four years old in 1960 (photo), 74
  music, 115
  rafting trip in 1982 (photo), 259
Streitwieser, William (brother)
  birth, 9
  education, 23
  recent picture (photo), 24
Stretching force constant, deuterium isotope effect, 64
Structural symbols, bonds, 177
Stuyvesant High School
  organic chemistry, 5
  Science Talent Search, 20–22
Suicide, first wife, effect on my chemistry, 1
Sulfuryl chloride, reaction with fluorene, 18
Superposition error, basis sets too small or unbalanced, 164
Synergy
  application to chemical life, 2
  definition, 2

T

Taft, Robert W., polar substituent constants, 60–61
Tatlow, J. Colin
  highly fluorinated bicyclo[2.2.1]heptanes, 117
  professor emeritus, Birmingham University (photo), 117
1,1′,4,4′-Tetra-*tert*-butyluranocene, $^1$H NMR spectra as function of temperature, 193f
Tetracation, proposed structure, 224
Tetrahydrofuran, carbanions, 125–143
Tetrapyridiniumcyclopropene
  development, 224–227
  twisted pyridinium rings, 231–232
Textbook
  flexibility, 243
  *Introduction to Organic Chemistry*, 238–243
Theoretical chemistry, chemical understanding, 2
Theory
  changes, 258
  organic chemistry papers, 243
THF, *See* Tetrahydrofuran
Thomas, T. Darrah, high acidity implied in electronic structure, 166
Thorocene
  f and d orbitals in ring–metal bonding, 202
  half-sandwich structure, 197
  $\chi\alpha$MO calculations, 208–209
Tinoco, Ignacio, at old chemistry building, January 1961 (photo), 47
Toczko, Glen, computer program for index, 240
Tolbert, Laren, Conference on Carbanion Chemistry, Ottawa, July 1989 (photo), 142
Toluene, microwave discharge products, 237

Torrence, Vivian, *Chemistry Imagined*, 215–216
Transition state
  computations of structures, 176
  effect of cation on structure, 171
  highly bent, 171
  ion-pair computations, 170
  ion-pair reactions, 170–176
  out-of-plane C–H bending motions, 65$f$
  proton exchange of methane with lithium amide, 172$f$
  proton transfer, 107
Transuranium elements, Lawrence Berkeley Laboratory, 195
Travel
  fluorine conference in Munich, 115–116
  Germany in 1976, 205
  Japan, 81
  total eclipse of the sun, 252
1,3,5-Trichlorobenzene
  chlorination of aniline in "outdoor facilities", 15–16
  Organic Specialties, 6–7
Triflate leaving group, solvolyses, 124–125
Triple-ion structure, electrostatic energy, 143
Tritium exchange, measurement, 97
Tritium exchange rates, Brønsted correlations, 105$f$
Tritium NMR, kinetics studies, 97
Truman, Harry, meeting at Westinghouse Science Talent Search, 20
Tsuno, Yuho
  during KISPOC I in 1982 (photo), 83
  visit to Japan, 81
Twisted pyridinium rings
  tetrapyridiniumcyclopropene, 231–232
  X-ray crystal structure, 234$f$
Typist, Army assignment, 26

U

Ugi, Ivar
  award nomination, 202

Ugi, Ivar—*Continued*
  hosted six-month stay in Munich in 1976 (photo), 204
University of California, Berkeley
  instructorship, 45
  Miller Research Professor for 1964–1965, 113
Uranocene
  discovery and development, 181–199
  f and d orbitals in ring–metal bonding, 202
  half-sandwich structure, 197
  meager reaction chemistry, 189
  substituted, 189
  X-ray crystal structure, 186$f$
  $\chi\alpha$MO calculations, 208–209
Uranocene derivative, NMR spectrum, 190
Uranocene synthesis, generating idea, 201
U.S. Army Medical Corps, service, 24–27

V

Van Sickle, Dale
  party at the Koch's in 1960 (photo), 91
  protodedeuteration of toluene-$\alpha$-$d$ with LiCHA, 87
Vogel, Emanuel
  discussing chemistry in Cologne in 1991 (photo), 184
  octalene synthesis, 183
Vorpagel, Eric, polarizations of sigma electrons in substituted benzenes, 165

W

Walker, Roger, uranocene hydrolysis rate, 216
Wang, Hsu-Kun
  instant translation, 198–199
  preparation of bridged uranocene, 197
  Second International Conference on Lanthanides and Actinides, Lisbon, April 1987 (photo), 199

Ward, Hal, magnetometer microwave discharge apparatus, 237
Warner, Chet, 1967 ACS Award in Petroleum Chemistry (photo), 248
Wasserman, Edel, Science Talent Search, 22
Waterman, Ken
  behavior of highly nucleophilic pyridine, 226
  X-ray crystallography, 225
Watson, Patricia, Second International Conference on Lanthanides and Actinides, Lisbon, April 1987 (photo), 199
Wave function
  development of concepts, 178
  lithiums $\pi$-bonded to propylidene dianion, 164
Weiss, Robert, fully pyridinium-substituted benzene, 236
Westheimer, Frank, chiral ethanol-1-$d$, 57
Westinghouse Science Talent Search, work with fluorene, 20–22
Wiberg, Kenneth
  Doering research group, 30
  Pauling Award (photo), 32
  remarkable organization, 30–31
  Marge, Vancouver, 1992 (photo), 32
Widom, Benjamin
  awards dinner in Las Vegas in 1982 (photo), 28
  physical chemistry laboratory, 28
  Stuyvesant High School, 22
Widower, adjustment period, 113
Wilcox, Charles, Science Talent Search, 22
Wiley, George, solvolysis reactions, 76
Wine cellar, home, 255
Wine tastings, 255–257
Winstein, Saul
  neighboring group effect, 58
  Ph.D. at UCLA, 15
Wittig, Georg, Roger Adams Award (photo), 80
Wolf, Al
  Doering research group, 30
  formation of $\beta$- and $\gamma$-fenchenes, 35–36
  in 1964 (photo), 33

Wolf, Al—*Continued*
  picnic excursion to Bear Mountain (photo), 38
  teaching assistant in advanced organic chemistry, 31–32
  temporary replacement teacher, 113
  Visiting Professor teaching organic chemistry, 112
Wolf, Helga, picnic excursion to Bear Mountain (photo), 38
Wolf, Richard A., computational quantum chemistry, 150
Women, proper place in society, 57
Woodward, Robert, evening group meetings, 43
Woodward–Hoffmann rules, pericyclic reactions, 77

Y

Yoshida, Norio, synthesis of thorium analog, 189
Young, Richard W.
  asymmetric induction, 36–37
  Doering research group, 30
  in 1950 (photo), 36
  role model, 32
Young, William R., hydrogen isotope exchange reactions, 95
Ytterbium, sandwich structure derivative, 200
Yukawa, Yasuhide
  during KISPOC I in 1982 (photo), 83
  physical organic chemistry (photo), 82
  tour around Osaka and Nara, 81

Z

Zalkin, Alan
  tenth anniversary of discovery of uranocene (photo), 185
  uranocene crystal structure determination, 185–186
Zeiss, Harold, solvolysis chemistry, 33–34
Zero-flux surface, spatial electron populations, 157
Ziegler, Gene R.
  kinetic acidity measurements, 121
  teaching and travel, 121
Zwanzig, Robert, Stuyvesant High School, 22

*Copy editing: Zeki Erim*
*Production: Margaret J. Brown and Donna Lucas*
*Indexing: Colleen P. Stamm*

*Production Manager: Cheryl J. Wurzbacher*

*Printed and bound by Maple Press, York, PA*

# Bestsellers from ACS Books

*The ACS Style Guide: A Manual for Authors and Editors*
Edited by Janet S. Dodd
264 pp; clothbound ISBN 0–8412–0917–0; paperback ISBN 0–8412–0943–X

*The Basics of Technical Communicating*
By B. Edward Cain
ACS Professional Reference Book; 198 pp;
clothbound ISBN 0–8412–1451–4; paperback ISBN 0–8412–1452–2

*Chemical Activities* (student and teacher editions)
By Christie L. Borgford and Lee R. Summerlin
330 pp; spiralbound ISBN 0–8412–1417–4; teacher ed. ISBN 0–8412–1416–6

*Chemical Demonstrations: A Sourcebook for Teachers,*
*Volumes 1 and 2,* Second Edition
Volume 1 by Lee R. Summerlin and James L. Ealy, Jr.;
Vol. 1, 198 pp; spiralbound ISBN 0–8412–1481–6;
Volume 2 by Lee R. Summerlin, Christie L. Borgford, and Julie B. Ealy
Vol. 2, 234 pp; spiralbound ISBN 0–8412–1535–9

*Chemistry and Crime: From Sherlock Holmes to Today's Courtroom*
Edited by Samuel M. Gerber
135 pp; clothbound ISBN 0–8412–0784–4; paperback ISBN 0–8412–0785–2

*Writing the Laboratory Notebook*
By Howard M. Kanare
145 pp; clothbound ISBN 0–8412–0906–5; paperback ISBN 0–8412–0933–2

*Developing a Chemical Hygiene Plan*
By Jay A. Young, Warren K. Kingsley, and George H. Wahl, Jr.
paperback ISBN 0–8412–1876–5

*Introduction to Microwave Sample Preparation: Theory and Practice*
Edited by H. M. Kingston and Lois B. Jassie
263 pp; clothbound ISBN 0–8412–1450–6

*Principles of Environmental Sampling*
Edited by Lawrence H. Keith
ACS Professional Reference Book; 458 pp;
clothbound ISBN 0–8412–1173–6; paperback ISBN 0–8412–1437–9

*Biotechnology and Materials Science: Chemistry for the Future*
Edited by Mary L. Good (Jacqueline K. Barton, Associate Editor)
135 pp; clothbound ISBN 0–8412–1472–7; paperback ISBN 0–8412–1473–5

For further information and a free catalog of ACS books, contact:
American Chemical Society
Distribution Office, Department 225
1155 16th Street, NW, Washington, DC 20036
Telephone 800–227–5558

# Highlights from ACS Books

*Good Laboratory Practice Standards: Applications for Field and Laboratory Studies*
Edited by Willa Y. Garner, Maureen S. Barge, and James P. Ussary
ACS Professional Reference Book; 572 pp; clothbound ISBN 0–8412–2192–8

*Silent Spring Revisited*
Edited by Gino J. Marco, Robert M. Hollingworth, and William Durham
214 pp; clothbound ISBN 0–8412–0980–4; paperback ISBN 0–8412–0981–2

*The Microkinetics of Heterogeneous Catalysis*
By James A. Dumesic, Dale F. Rudd, Luis M. Aparicio, James E. Rekoske, and Andrés A. Treviño
ACS Professional Reference Book; 316 pp; clothbound ISBN 0–8412–2214–2

*Helping Your Child Learn Science*
By Nancy Paulu with Margery Martin; Illustrated by Margaret Scott
58 pp; paperback ISBN 0–8412–2626–1

*Handbook of Chemical Property Estimation Methods*
By Warren J. Lyman, William F. Reehl, and David H. Rosenblatt
960 pp; clothbound ISBN 0–8412–1761–0

*Understanding Chemical Patents: A Guide for the Inventor*
By John T. Maynard and Howard M. Peters
184 pp; clothbound ISBN 0–8412–1997–4; paperback ISBN 0–8412–1998–2

*Spectroscopy of Polymers*
By Jack L. Koenig
ACS Professional Reference Book; 328 pp;
clothbound ISBN 0–8412–1904–4; paperback ISBN 0–8412–1924–9

*Harnessing Biotechnology for the 21st Century*
Edited by Michael R. Ladisch and Arindam Bose
Conference Proceedings Series; 612 pp;
clothbound ISBN 0–8412–2477–3

*From Caveman to Chemist: Circumstances and Achievements*
By Hugh W. Salzberg
300 pp; clothbound ISBN 0–8412–1786–6; paperback ISBN 0–8412–1787–4

*The Green Flame: Surviving Government Secrecy*
By Andrew Dequasie
300 pp; clothbound ISBN 0–8412–1857–9

For further information and a free catalog of ACS books, contact:
American Chemical Society
Distribution Office, Department 225
1155 16th Street, NW, Washington, DC 20036
Telephone 800–227–5558

# Other Titles from ACS Books

*The ACS Style Guide: A Manual for Authors and Editors*
Edited by Janet S. Dodd
264 pp; clothbound ISBN 0–8412–0917–0; paperback ISBN 0–8412–0943–X

*Enough for One Lifetime: Wallace Carothers, Inventor of Nylon*
By Matthew E. Hermes
History of Modern Chemical Sciences Series
350 pp; clothbound ISBN 0–8412–3331–4

*Roger Adams: Scientist and Statesman*
By D. Stanley Tarbell and Ann Tracy Tarbell
248 pp; clothbound ISBN 0–8412–0598–1

*From Small Organic Molecules to Large: A Century of Progress*
By Herman F. Mark
Profiles, Pathways, and Dreams Series
174 pp; clothbound ISBN 0–8412–1776–9

*American Chemists and Chemical Engineers*
Edited by Wyndham D. Miles
554 pp; clothbound ISBN 0–8412–0278–8

*Chemistry and Crime: From Sherlock Holmes to Today's Courtroom*
Edited by Samuel M. Gerber
135 pp; clothbound ISBN 0–8412–0784–4; paperback ISBN 0–8412–0785–2

*Nobel Laureates in Chemistry 1901–1992*
Edited by Laylin K. James
History of Modern Chemical Sciences Series
826 pp; clothbound ISBN 0–8412–2459–5; paperback ISBN 0–8412–2690–3

*Stalin's Captive: Nikolaus Riehl and the Soviet Race for the Bomb*
By Nikolaus Riehl and Frederick Seitz
History of Modern Chemical Sciences Series
240 pages; clothbound ISBN 0–8412–3310–1

*From Caveman to Chemist: Circumstances and Achievements*
By Hugh W. Salzberg
310 pp; clothbound ISBN 0–8412–1786–6; paperback ISBN 0–8412–1787–4

*Following the Trail of Light: A Scientific Odyssey*
By Melvin Calvin
Profiles, Pathways, and Dreams Series
200 pp; clothbound ISBN 0–8412–1828–5

*Seventy Years in Organic Chemistry*
By Tetsuo Nozoe
Profiles, Pathways, and Dreams Series
292 pp; clothbound ISBN 0–8412–1769–6

---

For further information and a free catalog of ACS books, contact:
American Chemical Society
Distribution Office, Department 225
1155 16th Street, NW, Washington, DC 20036
Telephone 800–227–5558